"I'd put my money on the sun and solar energy.
What a source of power!"

Thomas Edison, 1931

Jeremy Leggett is a social entrepreneur and climate-change campaigner. Originally an oil geologist, he founded the UK's leading solar company and is a director of the world's first private equity fund for renewables. In 2009 he became the first Hillary Laureate for Leadership in Climate Change Solutions. His previous books include *The Carbon War* and *Half Gone*.

The Solar Century

The past, present and world-changing future of solar energy

Edited by Jeremy Leggett

First published in Great Britain in 2009 by
GreenProfile, an imprint of Profile Books Ltd,
3a Exmouth House, Pine Street, London EC1R 0JH
www.profilebooks.com

1 3 5 7 9 10 8 6 4 2

The moral right of the author has been asserted

Printed and bound in Great Britain by Butler Tanner and Dennis Ltd,
Frome, Somerset

Typeset in DIN and Minion to an original design by Tim Godwin

A CIP catalogue record for this book is available
from the British Library

ISBN: 9781846688737

FSC
Mixed Sources
Product group from well-managed
forests and other controlled sources
Cert no. SGS-COC-005091
www.fsc.org
© 1996 Forest Stewardship Council

Acknowledgements

The writers of this book would like to thank everyone who supported its preparation. This doesn't just mean our immediate colleagues, but everyone in and around the solar industry who helped us along the way. The list, if we were to write it down, would be long. We sincerely hope we have done you, and the dream of a solar revolution, justice.

The writers and researchers, in alphabetical order, are: Seb Berry, Dan Davies, David Edwards, Gary Fowles, John Keane, Jeremy Leggett, Derry Newman, Nick Sireau, Alan South and Charlotte Webster. Aki Maruyama-Leggett and Olivia ten-Kate did additional research. Oliver Sylvester-Bradley oversaw the production of the book, and the designer was Tim Godwin. Duncan Clark was the editor at *Green*Profile and Jasmine Farsarakis, Dr Jenny Nelson, Professor Keith Barnham, Michael Priestnell, Michael Rogol and Sophie Orme provided additional editorial comment. The Ashden Awards for Sustainable Energy kindly supported the production financially.

The editor, Jeremy Leggett, bears most responsibility, because he edited the words of others as well as his own, and if they or he have made any mistakes, it is he who let them pass.

Information and dialogue

The writers have tried hard to make a clear distinction in this book between factual information and opinion. But facts can change as understanding evolves, opinions are very often subject to debate, and nobody is immune from making mistakes. For these reasons, and because debate is so urgently needed on how to power the future, we would welcome feedback on this book, negative or positive. To get in touch, email book@solarcentury.com

Technical terms, sources, and further reading

We have not assumed any more than a lay understanding of science among readers, and have tried to define terms unlikely to be widely understood as we go along in the text.

Any readers interested in the discussion about solar and poverty alleviation, should visit SolarAid's website, www.solar-aid.org, where there is news of the charity's programmes, information on how to get involved in them, and links to the sites of other development organisations.

People seeking more information on solar energy and the solar industry in general will find excellent coverage at the following websites: www.photon-magazine.com; www.solarbuzz.com; www.sunwindenergy.com; and www.nrel.gov.

Preface

by Achim Steiner

In the coming years, we may look back on the first decade of the twenty-first century as a time when governments, business and the global public began to glimpse and to foster a fundamental transition towards a Green Economy – one where the abundant forces of nature such as the sun, the Earth's hot rocks and the winds began to be harvested on truly transformational scales.

For as *The Solar Century* rightly points out, we really are on the verge of a solar revolution that, allied to energy efficiency and breakthroughs in other forms of renewables alongside technological improvements from battery capacity to smart grids, holds the solid promise of delivering clean, affordable and climate-friendly electricity for billions of people.

A revolution too in resource efficiency, rural electrification and rural development, education, innovation and new kinds of decent Green Jobs for the 1.3 billion unemployed or underemployed and the 500 million young people set to join the workforce over the next 10 years: in short a potentially significant contribution to the Millennium Development Goals.

Some may be tempted to say the solar promise has been made before, and has always been 'just on the horizon'. *The Solar Century*'s editor, Jeremy Leggett, has heard many of these critics. From his time as a geologist working for the oil industry to his role as founder and chairman of a leading UK solar company, Mr. Leggett is more qualified than most to assess what is hype and what is fact.

Claims in the book, such as building houses that use no direct or indirect fossil fuels is "surprisingly easy", may sit uncomfortably with some from government and industry, but the truth is that change can come quickly when governments, industry and individuals decide to act and when the market signals and mechanisms are in place.

The financial and economic crises facing the world at the time of this book's publication present an enormous opportunity for governments to forge a Global Green New Deal in which some of the $3 trillion-worth of stimulus packages is directed towards laying the foundations of energy and climate security en route to a truly sustainable 21st century.

As Victor Hugo once observed: "Nothing is so powerful as an idea whose time has come". For Mr. Leggett, this translates simply as 'seeing is believing'. When people buy a solar system, or live or work where one operates, he says, they quickly become passionate advocates of the solar approach, slimming down their electricity consumption and becoming more aware of energy issues in general.

As the world moves towards a new climate agreement, the United Nations Environment Programme firmly believes that this is the moment to take clean energy generation to a new forward-looking level and we are working across the UN system and in partnership with industry, the financial sector, governments and civil society to realize this aim.

I commend Mr Leggett and the authors for their valuable and inspiring work. I encourage you to read *The Solar Century* and become, like Mr Leggett, passionate advocates for a century of increasing solar, environmental and human progress.

Achim Steiner is UN Under-Secretary General and Executive Director of the United Nations Environment Programme (www.unep.org)

Editor's Foreword

In my youth, I became fascinated by geology, and immersed myself in learning about the natural rhythms of our planet over millions of years of geological time. Just over thirty years ago I joined a university faculty, at Imperial College of Science and Technology, to research Earth history. Imperial was an elite training school for the extractive industries, and like most of my colleagues I consulted all over the world for the oil industry, while training its future workforce. Some of the research I did was on the types of rock that create oil. Some was on the history of oceans. The more I discovered in the course of that oceanic research the more I appreciated how fragile and fundamentally changeable the planet is.

In the 1980s, growing numbers of scientists began to worry that the burning of fossil fuels – oil, coal and gas – would destabilise the relatively quiescent global climate that humankind has enjoyed as civilisation has evolved. When we burn these fuels – whether in power stations, cars, planes or buildings – greenhouse gases result. These gases trap heat in the Earth's thin atmosphere. At the time I was permitted a few lectures in which to point out the greenhouse problem. They were referred to by my faculty colleagues as 'Dr Leggett's liberal studies classes'.

Times change: at least in the appreciation of global problems, if not in the tackling of them. Today, believing that we can continue to pump billions of tonnes of radiatively active gases into the thin atmosphere each year without significant damage is a little akin to believing that the Earth is flat.

There are many ways we could stop greenhouse-gas emissions, if collectively we seriously wanted to. Most of them have to do with the way we use energy, wherein fossil-fuel burning can be replaced with a broad family of renewable-energy sources, plus energy efficiency and energy conservation. In 1989, I quit the university world to campaign for this switch to clean and efficient energy.

After seven years of lobbying at and around the international climate negotiations, and hard lessons in what some would call realpolitik, I came to the view that governments will struggle badly to summon a meaningful collective response to climate change. I could see some signs of potentially meaningful corporate responsibility emerging, but far too slowly, and the

signs were far too few. I decided to have a crack at setting up my very own microcosm for hope in the business world.

I founded a small company to try and install, on buildings, as much solar power as possible. I didn't do this because I considered solar to be a magic bullet. It is a single member of a big family of solutions, nothing more and nothing less. But it's a particularly neat solution, and a little magical, too. I and many other solar fans believe it can be the backbone of energy supply in a sane and survivable future society.

I called the company Solarcentury because that is what I believe the twenty-first century has to be. I think of this book as a progress report, nearly one decade into the century, on the viability of that vision.

Jeremy Leggett,
London, April 2009

Contents

Introduction   *xii*

Chapter 1 - The Big Picture   *1*
Energy in our Solar System

Chapter 2 - Crunch Time   *17*
Why We Need a Clean-Energy Revolution

Chapter 3 - The Energy Revolution   *35*
How Solarisation Can Power the World

Chapter 4 - Solar Tech   *57*
How Solar Technologies Harvest Energy From the Sun

Chapter 5 - The Solar Revolution   *97*
Visions of Solar in Action Today and Tomorrow

Chapter 6 - Making it happen   *141*
How to Make the Solar Century a Reality

Endnotes & References   *157*
Index   *163*

Introduction

As this book goes to press, it is becoming increasingly clear that a great global programme of energy reform is needed. Nothing else can cut human greenhouse-gas emissions deeply enough to head off the worst impacts of climate change. At the same time, it is also becoming evident that the global oil and gas industries will soon struggle to meet demand for their demonstrably finite products, even in the face of global recession. Growing numbers within and around the oil industry say oil production is near its peak, and is in danger of soon beginning a decline. Many refute this peak-oil analysis still, but among these doubters there are many who do accept that the age of easy oil is over, and/or that the geopolitics of oil- and gas-supply are such that most nations face profound energy-security threats. On top of the climate and energy-security problems sits a major financial crisis, which poses many threats that compound the other two, not least in the perceived added difficulties of financing clean energy alternatives that abate the climate crisis while buying time to insure against an early peak in oil production.

These three dilemmas – of climate, energy and finance – we can think of as a 'triple crunch'. Each requires a collective response for which there are no equivalents in the history of human civilisation. Together, they pose humankind with compound challenges that will probably determine whether or not civilisation can survive.

Thankfully, many tools are available to help counter these threats. This book is about one of the most important: solar energy.

Just how important a tool is solar? And can renewable-energy sources really save us from climate change and energy insecurity? To give a flavour of the kinds of arguments we will make in answering these questions, let's briefly touch on how solar and renewables can fit into society at various levels, considering in ascending scale the home, the town, the city, the state and the nation.

Building houses that use no fossil fuels, directly or indirectly, is surprisingly easy, and growing numbers of such houses are appearing, both in cooler high latitudes as well as the sunbelt. One American home, in Cape Cod, Massachusetts, looks no different from any typical New England house, except in one way. Solar-photovoltaic panels (those which produce electricity from sunlight) and solar-thermal collectors (panels that produce hot water) sit on its roof. They generate much of the electricity and heating the residents need. The rest is provided by a geothermal heat pump that uses groundwater to heat or cool the home. No fires are needed in winter, no air conditioning is required in summer. The house cost 23% more to build with these and other green features than in the usual way. A 23% premium seems high until one considers that it can be

Solar pioneer Jeffrey Rogers, outside his Cape Cod house.
(Image: Alice Shafer)

recouped, net of state and federal incentives, within five years. Thereafter the homeowner will be saving increasing amounts of money every year, for many years.[1] Later in the book we will look further at solar economics of this kind, and argue that they make complete sense today: something which is not the general perception.

Such an argument applies even more clearly to the developing world. The vast majority of homes in the developing world depend on kerosene for light at night. It is easy to adapt kerosene lanterns to use solar cells and batteries, and recouping the small investment on the equipment needed to do that – in terms of kerosene bills avoided – is measured in months. With the right collective will, it would be easy to replace kerosene lanterns with solar ones in every home in the developing world.

These two examples, we believe, are microcosms of what can be done in residences. Provided energy efficiency is maximised alongside solar supply, it is possible in principle to swap fossil fuels with solar in the vast majority of homes across the world. Though we could do this, we do not actually need to do it in every home. After all, we have many other members of the renewable-energy technology family to mix and match with. (One reason why we chose the New England example above, and not a 100% solar-powered home.)

At the scale of a town, Woking in the UK offers a good example of the mix-and-match approach. Using a mix of combined-heat-and-power (CHP) plants (which use the waste heat from electricity generation to

produce hot water), solar photovoltaics (PV) and some other renewable technologies on local electricity grids, the council in Woking has cut its carbon dioxide emissions deeply since the early 1990s. By using local electricity grids, the town has also provided its residents with the security of knowing that, if the national grid goes down, the lights stay on in their homes, the water from the taps will still be hot, and the heating will still come on in winter. Woking burns small amounts of natural gas in the CHP plants, but this could in principle be gas produced sustainably from food wastes and other sources of biomass. Woking council's greenhouse-gas target is zero emissions, and there is no doubt they can hit it.

If small towns in cloudy England can do this, why not all other towns?

Cynics might respond: 'Because humans don't like change, and anyway you couldn't do it for a whole county or state.'

Let us set aside the issue of change for a moment, and consider the proposition about the county or state. The Spanish state of Castile-La Mancha has already hit 40% of renewables in its energy mix en route to a goal of 100% by 2012. According to the plan, solar PV would be 15% of the mix, alongside wind, biomass and 20% increased efficiency. The plan is captured in a regional law.

If a state in Spain can do this, why not every state in every nation?

At the scale of the nation, as we'll see later in the book, a growing number of studies and case histories show that entire countries can be powered by renewables. One example of a country which is targeting this is Portugal. In less than three years, Portugal has quadrupled its wind power, trebled its hydropower, built a 'solar farm' photovoltaic (PV) power plant twice the size of any other in the world, and opened the world's first commercial wave-power station.

If Portugal thinks it can do it, why doesn't every other nation try?

Cynics might respond: 'Because humans don't like change, and anyway you couldn't do it for a major greenhouse emitter, like China, or Russia, or America.'

Many people in the renewables industry are confident that 100% renewables energy supply for the entire world is possible, and sooner than most energy pundits would acknowledge. For example, the overwhelming consensus of participants at the Tenth Forum on Sustainable Energy, held in Barcelona in April 2008, was that the world could be 100% renewable-powered within twenty to forty years.

But let us pause and cede the cynics some ground, temporarily. The greenhouse-gas situation does not look encouraging as we consider the great powers today. China builds giant coal-fired power plants apace. Russia uses its domestic oil and gas wealth to flex muscles aggressively on the international stage. America and Europe talk of reducing emissions and weaning themselves off fossil fuels, but they have so far achieved relatively little progress towards these aims. That bloc of sometimes individually quite encouraging nations, Europe, has a dismal collective record on greenhouse-gas emissions.

But there are still plenty of grounds for optimism. If we can produce the right leadership, at all levels of society, we can overcome the human resistance to change and unlock the power of

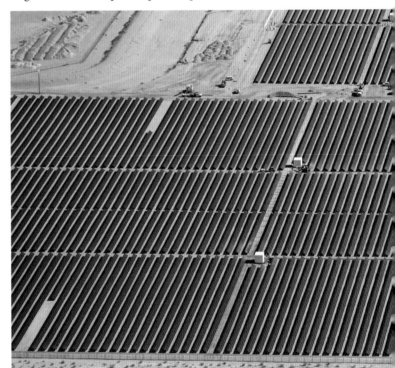

A solar photovoltaic farm in Nevada, USA
(Image: Sempra Energy/First Solar)

solar and the rest of its relatives in the clean-technology family. If good leaders lead, others tend to follow. Certainly this has been the case where individuals champion solar effectively, as German parliamentarian Hermann Scheer has done. Scheer has argued passionately and persuasively over the years that renewables really can run economies. Partly as a result, the German government has introduced policies that have grown what is now the biggest solar-photovoltaic market in the world.

Others have followed where cities champion solar effectively, as with Rizao in China. In parts of Rizao, 99% of households use solar water heaters. Other Chinese cities have emulated this, and China's solar hot-water market has become bigger than those of all the rest of the world combined.

Others have followed where governments lead effectively, as they have in Japan and Germany, where forward-looking policies encouraged Japanese and German companies to leap into the embryonic solar industry in the 1990s and early 2000s. These countries rapidly dominated export markets. Seeing this, other companies and countries have copied them, in China's case spawning new solar corporations, some of which have since outstripped their Japanese counterparts.

Investment in all renewables globally totaled nearly $150 billion in 2007, according to the United Nations Environment Programme. This is more than a tenth of total annual global investment in energy (around $1.3 trillion in 2007), whereas modern renewable energy provides little more than two percent of total world energy. Moreover, this investment was growing fast at the time of the financial crash: the 2007 figure was a five fold increase on 2004's $33 billion. The fastest growing sector for investment was solar photovoltaics, with more than $28 billion of new capital in 2007.[2] Investor interest in the sector is continuing through the downturn, with more than 50% of all venture capital invested in cleantech invested in photovoltaics in 2008.[3]

Driven by this rising investment, the global solar-photovoltaic market grew fully 69% in 2007 and has averaged 44% growth since 1999: faster than laptops, mobile phones, digital cameras and many of the other electronic goods that consumers know so well today. Things have been little different in solar thermal, where the European market grew 47% in 2006, doubling in less than three years. The drivers for this solar-thermal growth include national and city government leadership in market-support schemes. The fact that Russia turned off Europe's gas supply briefly in January 2006, 2008 and 2009 also made a difference, engendering as it did widespread nervousness about relying on energy imports in the future. On this kind of evidence, solar-photovoltaic and solar-thermal markets seem set to continue growing fast, notwithstanding the inevitable setbacks that all markets are suffering in the current financial crisis.

As anyone who has worked in the solar industry knows, there's another reason to believe that we're on the verge of a solar revolution. It's what one might call the 'seeing is believing' effect. Time and time again, we find that, when people buy a solar system, or live or work where one operates, something interesting happens. A light bulb tends to goes on, so to speak, in that allegedly change-resistant human brain. People quickly become passionate advocates of the solar approach; they slim down their electricity consumption, and become more aware of energy issues in general.

The challenge is simple for those of us who believe that history isn't destiny, that fundamental change for the better is possible, and that even countries like China, Russia and America can power their economies with renewables. We must take this 'seeing is believing' effect and proliferate it until it reaches a tipping point.

This book, we the writers hope, will play a small part in meeting that challenge. We seek to encourage readers to believe that society can turn its back on the fossil-fuel century just past; that people, communities, businesses and governments together can stage a rapid but managed retreat from fossil-fuel dependence; that solar technologies can play a key role in this process; and that much social good can come about as a result of the continuing rapid growth of solar markets. In short, we hope to convince readers that the twenty-first century can become 'the solar century', and – if it does – that we can reasonably expect to avoid the worst impacts of global warming, help make energy crises a thing of the past, create many jobs, and generally improve quality of life along the way.

Chapter 1
THE BIG PICTURE

Energy in our Solar System

Cosmologists tell us that the real beginning of the solar story is the Big Bang, the cataclysmic event around 13–14 billion years ago when matter, time, energy and space all came into being. A faint heat radiation emanating from all points of the sky is among the evidence that persuades physicists that the universe can only have been generated as an instantaneous fireball, and that it has been expanding ever since. After the Big Bang, strands of hydrogen and helium gas coalesced under the influence of gravity, over a period of 1 or 2 billion years, to form galaxies, of which there are around 100 billion in the universe. Within each galaxy, matter continued to concentrate, forming the first stars. Within the stars, as matter coalesced further under the draw of gravity, hydrogen and helium atoms became capable of fusion, so releasing vast amounts of heat and light. These stars are the specks that adorn our night sky, each of them a hot ball of gas, mostly made of hydrogen and helium, transmitting their energy through space to us at the speed of light.

The sun as a star

One of these galaxies, a mere speck within the universe, was the Milky Way. Within this galaxy, a clump of gas and dust began coalescing about 4.5 billion years ago into what would become our solar system, itself a mere speck within the Milky Way. It consists of a central sun, nine orbiting planets and all the leftover bits known as asteroids and comets.

Today our sun contains 99% of the solar system's mass. Around 1.4 million kilometres in diameter, it is more than a million times bigger than the Earth, which orbits it at a distance of roughly 150 million kilometres, third in sequence among the nine planets. Our sun is of a type with

Earth.
(Image: NASA/courtesy of nasaimages.org)

only medium mass. Counterintuitively, suns of higher mass may live for only a few million years. Ours, so astrophysicists calculate, will burn for around another 4,500 million years.

Temperatures in the centre of the sun reach about 15 million degrees Celsius. The surface is at around 5,700 degrees Celsius – still unimaginably hot when you consider that lead boils into a gas at less than 2,000 degrees.

In the sun's super-dense core, hydrogen atoms fuse to form helium at a rate of 700 million tonnes per second. Environmentalists like to say they are in favour of nuclear power provided there is just one massive fusion reactor getting on for 100 million miles away from the planet. That one they can't escape, and nor would they want to.

Orbiting satellites have measured the vast flow of energy from the sun at the top of the Earth's atmosphere since 1978. There, the irradiance (a measure of energy-flow per unit of area) averages 1,368 watts per square metre. (See the box on p4 for definition of a watt, and other units of power and energy.) Half is visible light, almost half is infrared (IR) and a tiny but significant amount is ultraviolet (UV). This radiation is of a mix and intensity that has allowed life on Earth to flourish, while disallowing it on each of the other eight planets in the solar system.

Visible radiation allows photosynthesis, a crucial process by which most plants generate their own food, and upon which most animals depend, whether directly as herbivores or indirectly as carnivores. Infrared radiation heats the oceans, the land and the atmosphere to tolerable levels. It does so unevenly across the planet, thereby driving weather and the global climate.

Through the last 10,000 years or so, the relative stability of this climate has allowed crops to be grown with a fairly high predictability of success, and water supplies to be relied on with a relatively high degree of confidence. This in turn has allowed civilisation to evolve. With the advent of massive greenhouse-gas emissions, all that has been placed in jeopardy.

Let us also note that the sun probably played a crucial role in the creation of life on Earth. Scientists have known since 1953 that subjecting a mixture of ammonia, methane, hydrogen and water to an electrical discharge akin to lightning can cause organic compounds to form in a 'primordial soup', including most of the amino acids needed to build proteins, the building blocks of life. More recently, it has been discovered that irradiation of the same mixture by UV light promotes the formation of a vital energy-carrying molecule known as ATP, which is found in the cells of all living things. We have a lot to thank the sun for.

Our sun.
(Image: Courtesy of SOHO/EIT consortium. SOHO is a project of international cooperation between ESA and NASA.)

Power, energy, and their measurement

A **watt** is a measure of power, the rate at which energy flows (i.e. is turned from one form of energy into another).[1] A set of appliances might consume 1,000 watts, or a **kilowatt**, at any time. One way to measure energy is in terms of power used in a given period. So if we run the appliances for an hour, all on at the same time, they would need a **kilowatt-hour** of energy. Run them for an entire day and you would need 24 kilowatt-hours.

A small solar-photovoltaic roof might generate a kilowatt when incoming sun is at its strongest, but of course less when clouds shut out direct sunlight. Over the course of the year, in a cloudy country like Britain, photovoltaic panels of 1 kilowatt at their peak power (peak power is the way they are rated on the label) might generate something like 850 kilowatt-hours of electricity. In Mediterranean countries the same system might generate 1,500.

Power units scale up as follows. A million watts is a **megawatt**. Individual renewable-energy installations are most usually measured in megawatts of power. A billion watts is a **gigawatt**. Big central power plants are most usually measured in gigawatts. A thousand billion watts is a **terawatt**. Global and national power demand is usually measured in terawatts. Global and national energy demand is often expressed in **terawatt-hours per year**.

All these units are frequently abbreviated. For example, megawatt is often written **MW**, while kilowatt-hour is abbreviated to **kWh**.

Seasons and insolation

The Earth's 365.25-day orbit of the sun is not circular, but slightly elliptical. The planet also spins as it orbits, of course, which is what gives us night and day. It does so with a permanent tilt, at 23.5 degrees from vertical. This is what gives us the seasons. At the summer solstice in June, the northern hemisphere is tilted towards the sun, the north pole is sunlit all day, and the south pole is in permanent darkness. The situation is reversed by the time of the winter solstice in December. The sun's tilt also defines the tropics, with the sun being overhead at midday on the tropic of Cancer (23.5 degrees N) in June and at the tropic of Capricorn (23.5 degrees S) in December. Solar heating is greatest in the tropics. The further one travels towards the poles, the more oblique the sunlight, and the thicker the amount of atmosphere the photons have to travel through.

As we've seen, the irradiance at the top of the atmosphere averages 1,368 watts per square metre. As this radiation passes through the atmosphere it is reduced by reflection, absorption and scattering. Ozone, water vapour

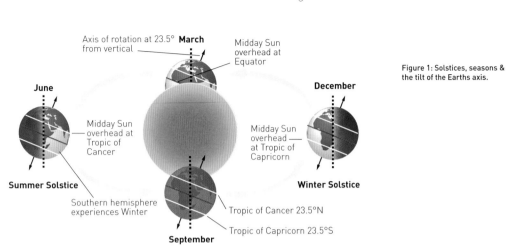

Axis of rotation at 23.5° **March**
from vertical

Midday Sun
overhead at
Equator

June

Midday Sun
overhead at
Tropic of
Cancer

Summer Solstice

Southern hemisphere
experiences Winter

December

Midday Sun
overhead —
at Tropic of
Capricorn

Winter Solstice

Tropic of Cancer 23.5°N

Tropic of Capricorn 23.5°S

September

Figure 1: Solstices, seasons &
the tilt of the Earths axis.

and carbon dioxide are among the gases that absorb radiation. Particles of dust and pollution are among those that scatter it. By the time the sun's energy reaches the Earth's surface, a midday maximum of around 1,000 watts per square metre remains. Reflected radiation from sunlight bouncing off clouds (diffuse radiation) can increase this figure, meaning that sunny days that are partly cloudy are best for solar energy.

If we add up the energy content of the radiation throughout the year, we arrive at a figure for annual global radiation in kilowatt-hours per square metre. This figure varies with location, as Figure 2 shows. Some equatorial regions can exceed 2,300 kilowatt-hours per square metre. Cloudy Germany, by contrast, averages little more than 1,000. This difference has not stopped Germany being the fastest-growing domestic solar-photovoltaic market in recent years. When people think that solar energy is technology for sunny countries only, as they often do, they are quite wrong.

Figure 2: Worldwide distribution of annual solar irradiance. Units are kilowatt-hours per square metre.

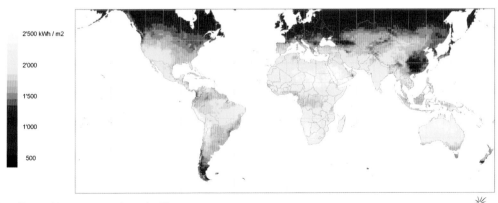

2'500 kWh / m2

2'000

1'500

1'000

500

Source: Meteonorm 6.0 (www.meteonorm.com); uncertainty 10%
Period: 1981 - 2000; grid cell size: 1°

June 2008

Sun worshippers

Human societies have long recognised the crucial, life-giving role of the sun, which probably explains why they have afforded it a central place in their religions and mythologies over the centuries. As any traveller who has visited the ruins along the Nile will know, the sun god, Ra, was the principal deity for the ancient Egyptians, who gathered in temples to celebrate the sun's rebirth each morning. The sun god took different forms. The visible disc of the sun the Egyptians called Aten, believing it to be the body or eye of Ra. One pharaoh, Akhenaten, even created a monotheistic religion built around Aten.

Many ancient societies in the Americas worshipped the sun, too. The Aztecs (AD 1325–1521) believed that there had been five worlds, or eras, each with its own sun. The previous four had ended in cataclysm. They persuaded themselves that, in order to ward off another cataclysm, human sacrifices were needed as offerings to the fifth sun. An Aztec calendar stone found on the site of modern Mexico City in 1790 places the sun god, Tonatiuh, right in the centre of a portrayal of their world.

Mexican sun stone, showing the sun god, Tonatiuh.

In ancient Europe, Greeks and Romans worshipped one or more sun gods. The Greek myths gave the sun the character of a Titan, Helios, who rode a chariot across the sky to cause day. They also associated the sun with the god Apollo. The Romans had their own version of Helios, called Sol. Sol Invictus (the undefeated sun) was a term they applied to several solar deities. The birthday of Sol Invictus was celebrated as a festival on 25 December from as early as the first century AD, later becoming a Christian ceremony with an altogether different significance.

The sun has featured prominently in Asian and Middle Eastern religion, too. In Asia, the sun has long figured in the deities of ancient China, though never as supreme deity. In ancient Japan, the sun goddess, Amaterasu, was one of the most prominent deities. She was once worshipped in the imperial palace itself, as a sacred ancestor of the royal family. The main shrine to her is the most important Shinto shrine in modern Japan. In Hinduism, the chief solar deity is called Surya, and a ritual performed by some Hindus focuses on worship of the sun. Hindu myths from India refer to the sun as a king riding on seven horses, representing the colours of the visible solar spectrum. Islamic shariah, by contrast, forbade worship at sunrise and sunset expressly to refute the sun's divinity.

Ancient civilisations also built their calendar around the sun. The oldest sundial, in southern Egypt, is 6,000 years old. Stonehenge, a 4,000-year-old British monument of the same vintage as the Egyptian pyramids, is essentially a solar observatory, consisting of enormous standing stones. Erected by building methods that remain a mystery, in an era with no written history, it is aligned with the rising sun at the solstice.

Today, the sun may not be worshipped as widely as it once was, but with a liveable future for humankind threatened by global warming the potential for solar-powering economies billions of years hence has given our star a whole new type of significance.

Stonehenge - the earliest example of British solar architecture?

Harnessing the sun

The current global power requirement – for all types of energy use including transport – is 13 terawatts, growing to about 20 terawatts in 2020 unless we get serious about energy efficiency. The amount of sunlight falling on the planet at any one time is around 120,000 terawatts. In other words, the amount of solar irradiance available for capture is well over 9,000 times the global power requirement. Stated another way, if we capture a tiny fraction of 1% of that solar irradiance we could provide more power than the world currently needs. Add a little energy efficiency and some of the other renewables, and we could do it with ease, in principle.

There are five main ways to harness that tiny fraction of total solar radiation. We will introduce these very briefly here, and consider them all in more detail later in the book.

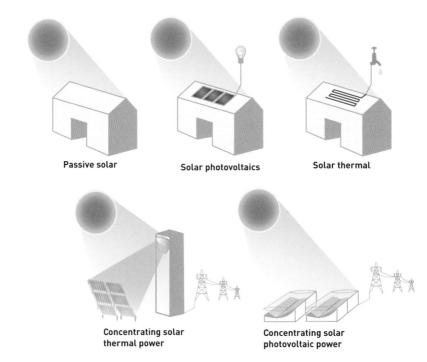

Passive solar **Solar photovoltaics** **Solar thermal**

Concentrating solar thermal power **Concentrating solar photovoltaic power**

Figure 3: The basic different types of solar power.

Passive solar

Architects have designed houses since before the time of the ancient Greeks to maximise the use of sunlight for heating in winter, while banishing much of the sun's heat in summer. Since the sun is high in the summer and low in the winter, this is relatively easy to do with a well-designed portico. Twelfth-century cliff dwellers in Colorado are known to have designed adobe houses in this way. With modern technology, other tactics are possible, such as movable louvres, or adjustable smart windows that reflect heat in summer but retain it in winter.

Passive solar - south facing building design.
(Image: Mischa Hewitt, www.lowcarbon.co.uk)

Solar photovoltaics

This technology makes use of the photovoltaic effect, whereby semiconductors become excited when exposed to light, and release electrons. These can be collected by wires attached to the semiconductor, and carried away as electricity. The electricity is of the direct-current type. As will be explained on p.40, in most usage today, direct current needs to be converted to alternating current. The piece of electronic equipment needed to do this is called an inverter.

Solar photovoltaic cells.

The basic building block of a solar-photovoltaic system is a solar cell, generally a postcard-sized slab of semiconducting material – usually made of silicon. These cells are assembled into modules ranging from book-size to door-size. The modules can be arrayed in open spaces, either in a fixed position or on trackers – computer-controlled mounting systems designed to follow the sun so as to maximise irradiance of the cell.

Solar PV systems are grid-connected or stand-alone. Grid-connected systems are either directly connected to the public electricity grid, or connected to it via the house grid. Stand-alone systems usually come with a storage device such as batteries. Inverter technology has advanced to the point that grid-tied solar systems can instantly be switched into stand-alone mode, should the owner wish.

A solar-thermal
collector in the UK.

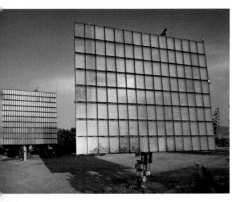

Solar photovoltaic
concentrator arrays.
(Image: Concentrix)

Solar-thermal

Solar-thermal technologies makes use of solar radiation to heat fluids of different kinds, in a number of different ways. Simple solar-thermal collectors use sunlight directly to heat fluids in metal tubes or flat, hollow metallic plates. Evacuating air from glass tubes makes heating of fluid within the tubes easier. The hot water is then piped away to a storage tank, either for use directly or to be further heated by other means.

Concentrating solar photovoltaic power

Lenses can be used to concentrate solar radiation onto photovoltaic cells. This focusing of energy onto the cell increases the efficiency of the device. Mirrors are also cheaper than photovoltaic cells, so by using less photovoltaic material overall system costs can in principle be reduced.

Concentrating solar-thermal power

Mirrors can be used to focus solar radiation on a tube, intensifying solar-thermal heating. Such processes can be applied at power-station scale, where fluids heated by solar radiation to very high temperatures are used to drive turbines. This can generate electricity on a scale of great interest to traditional electricity utilities, as we will see.

The five basic techniques can be used in combination. For example, a solar roof can have both solar-photovoltaic and solar-thermal roof tiles. The solar techniques can also be used with other members of the renewables family, and also with fossil fuels. For example, in cooler climates a solar-thermal collector might be combined with a small, highly efficient gas boiler that uses either biogas or natural gas. Such combinations are known as hybrid systems. The options are many.

Solar-thermal concentrators in the Nevada desert.

The sun and renewables

In a sense, most forms of energy could be described as being 'solar'. Hydroelectric power is entirely dependent upon the sun to drive the hydrological cycle that causes the precipitation that drives the rivers. The wind used by wind turbines blows as a result of differences in atmospheric pressure caused, primarily, by the uneven distribution of sunlight around the planet. The wind in turn pushes the waves that drive wave-power devices. The sun is integral to photosynthesis, in that solar energy acts on the pigment chlorophyll in plant material to create chemical fuel for plant growth, hence wood and other biomass energy sources. Even tidal energy has a solar component. Though the moon's gravitational pull does most of the work, the sun's gravitational influence is around half that of the moon's. Only geothermal power (the use of natural heat from the ground) and nuclear power have no association at all with the sun.

Clearly, we can speak of the deployment of a mixture of any renewables – other than geothermal – as a process of 'solarisation'.

The sun and fossil fuels

Fossil fuels are also very much associated with the sun. Indeed, one can justifiably think of coal, oil and gas as forms of fossilised solar energy. Here is why.

The planet's early atmosphere was rich in carbon dioxide and, though we don't know exactly how life first evolved, we do know from the fossil record that the earliest organisms were primitive marine photosynthesisers. They used chlorophyll in their cells, stimulated by sunlight, to combine carbon dioxide and water and form carbohydrates, giving off oxygen as they did so. They then burned the carbohydrate as fuel to produce energy for their cellular processes, giving off carbon dioxide and water again. These photosynthesisers appeared early in the Earth's history, just a few 100 million years after the creation of the planet and the solar system around 4.5 billion years ago. Palaeobiologists believe that, over several billion years, oxygen slowly built up in the atmosphere due to the success of the simple photosynthesisers, and carbon dioxide concentrations fell. Eventually life emerged on land, about 460 million years ago, in part due to these changes to the atmosphere's make-up. All this had taken a very long time. Multicellular organisms didn't appear in the seas until just over 600 million years ago, and flourished from that time onwards, barring several phases of mass extinctions variously attributed to meteorite impacts, huge volcanic episodes and changing alignments of the Earth's tectonic plates.

Forests first appeared around 360 million years ago, and within another 40 million years, in a period of time aptly called the Carboniferous, they were thick on the Earth. Mass burials of trees and other plants from these forests in oxygen-free environments over millions of years eventually resulted in the formation of coal seams, as sediment layers built up on top of them, exerting the pressures necessary to compact them into rock form. Every day in the world's coal mines, miners see fossil impressions of the leaves of the giant plants from which coal is made. Each of the leaves trapped solar energy. Each turned it into simple carbohydrate fuel for life. Each leaf and tree thrived as a result, in just the same way as the algae that created the atmosphere allowing plants to live on land in the first place. The mass burials of Carboniferous forests, and forests from other coal-forming times, trapped a vast quantity of that photosynthetic fuel, unoxidised, underground. Anyone who has seen coal shovelled into a fire gets a sense of how much.

Massive algal blooms in ancient oceans - much bigger than this bloom in the Bay of Biscay - gave rise to the organic matter in oil.
(Image: ESA)

While the forests grew on land, the marine algae flourished in the oceans, as they do to this day, sitting at the bottom of the food chain that humans rely on. They too experienced phases of mass burial, particularly around 150 and 90 million years ago – during the long tenure of dinosaurs on the planet. The sea floors where the burials took place were also devoid of oxygen, meaning the carbon wasn't oxidised.

The mass burials of algae at sea had a key difference from the mass burials of forests on land. The type of organic matter found in marine algae meant that, as the carbon was buried and heated up, it formed a liquid rather than a solid carbon rock. This liquid was oil. Some of the oil was able to migrate into porous rock formations and become trapped underground. The places where this happened are called oilfields. Oil forms at particular depths in rock columns. Deeper than around 15,000 feet, the temperatures reach more than 150 degrees Celsius, and the hydrocarbon liquid is turned into a gas. This is one way natural gas can form. Another is when carbon from plant matter other than algal becomes trapped under sediment without having been oxidised. Natural gas can migrate in much the same way as oil, entering rocks with pores or fractures, becoming trapped in some places, which later become gas fields.

In the case of all three fossil fuels, the basic story is the same. Sunlight meets carbon in conditions of life, bringing about energy for further life, and energy that can be trapped in vast quantities when the conditions are right for unimaginable numbers of plants – whether trees, algae, or other – to die en masse over a long period and not be oxidised. Rock cover does the rest, burying the whole of this fossilised-sunlight story, and trapping it undisturbed for millions of years. Until, that is, some species comes along clever enough to mine or drill it, and burn it, so releasing the trapped energy – and carbon – into Earth's thin atmosphere, effectively in a geological instant.

One portrayal of how much fossil fuel remains underground waiting to be burned, shown in Figure 4, compares the volumes of remaining coal, oil and gas with total annual energy consumption by humankind, plus available solar radiation at the Earth's surface. The available solar energy dwarfs the fossil fuels. We wouldn't need to use any solar energy, of course, until the finite fossil fuels had run out, if it were safe to burn the remaining coal, oil and gas.

But of course, it isn't.[2]

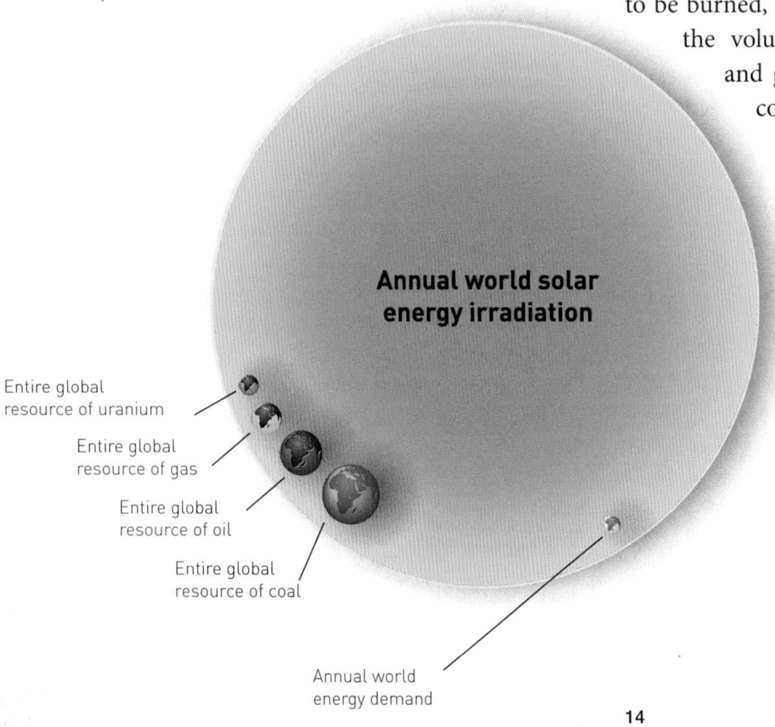

Figure 4: Approximate energy content of solar radiation reaching the earth's surface annually compared to global energy use. (based on BMWi 2000 - Planning and installing PV systems - Earthscan 2008)

Annual world solar energy irradiation

Entire global resource of uranium

Entire global resource of gas

Entire global resource of oil

Entire global resource of coal

Annual world energy demand

Chapter 2
CRUNCH TIME

Why We Need a Clean-Energy Revolution

This chapter considers the problems of climate change and energy security in a little more detail, before we move to solutions in Chapter 3. But first, let us briefly review the state of play in world energy reserves and consumption. This is essential background for what follows.

Fossil fuels: global reserves and resources

Society has relied upon the same main energy sources for well over a century. Oil was first discovered as long ago as 1859 and was in widespread use by the end of the nineteenth century. Today more than 4 billion tonnes of oil are consumed each year, meeting over 30% of world primary energy demand. Large-scale coal burning began even longer ago, in the Industrial Revolution. Today, coal meets more than a quarter of global primary energy demand: the energy equivalent of 3 billion tonnes of oil. (Oil-equivalent is a standard measure that makes it easier to compare different energy sources.) Gas didn't come into widespread use until the late twentieth century. Today it provides around 2.4 billion tonnes of oil-equivalent, representing more than 20% of world primary energy demand.

Renewables contribute around 13% to global energy consumption today. But of this, most involves unsustainable uses of wood and hydropower. (Many large hydropower projects cause significant environmental and social problems, and cannot be thought of as sustainable.) The proportion of what we might think of as 'green new renewables' (wind and solar, mainly) is little more than 2%.[1] Nuclear, after half a century of development, contributes only 6% to global energy demand.

Clearly, the world is very dependent on fossil fuels. It will be extremely difficult to reduce their roughly 80% share of primary-energy supply. If we have to reach zero carbon emissions from the energy sector in the decades ahead, as many policymakers now believe, it will be uniquely challenging.

The vertical columns in Figure 5 show the size and national ownership of the fossil-fuel reserves, as the energy industry reports them. 'Reserves' are deposits of coal, gas or oil that have been mapped underground by geologists, and can be extracted in a manner deemed conventionally 'economic' – that is, they can be mined or drilled and used or sold at a financial profit (regardless of the costs to the environment and society). Geologists also recognise 'resources', which are known reserves plus other deposits that have yet to be discovered or mapped, and which might one day be deemed economic. Resources are not shown in Figure 5, but in the case of coal and unconventional oil and gas such as tar sands and gas

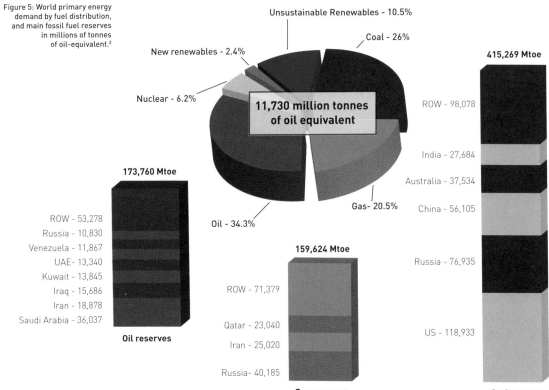

Figure 5: World primary energy demand by fuel distribution, and main fossil fuel reserves in millions of tonnes of oil-equivalent.[2]

hydrates (ice-like deposits of natural gas that build up in sediments rich in organic matter, especially in the Arctic) they far exceed reserves.

Global fossil-fuel reserves amount to 748,000 million tonnes of oil-equivalent, or almost three-quarters of a trillion tonnes. And that's before you add in the resources that energy companies are still exploring for. More and more people in and around the oil industry believe that the reserves numbers are overinflated, perhaps especially for oil, a point we return to below. But even if they are inflated by a factor of two, there is enough fossil fuel left to tip the world into runaway climate change many times over.

On top of the carbon problem comes a set of energy-security problems because some countries are so well endowed with fossil fuels while most of the rest have little. Almost 70% of oil reserves are in the hands of the biggest six OPEC (Organisation of Petroleum Exporting Countries) producers, plus Russia. Well over half the gas reserves are in just three countries: Russia, Iran and Qatar. History suggests there is plenty of scope for oil- and gas-producing countries to fall out with consumer countries, and vice versa. For the oil- and gas-consuming nations, the eggs really are in a very few baskets.

Of the coal reserves, which make up well over half the global fossil-fuel total, more than three-quarters are in five countries. These – the US, Russia, China, Australia and India, in order of the size of their reserves – have been among the nations most resistant to substantive multilateral agreement on efforts to cut global greenhouse-gas emissions, and thus fossil-fuel burning.

In terms of the location of resources, as opposed to reserves, the story is little different. Figure 4 in Chapter 1 gives one depiction of the relative ultimate size of resources. Coal is the biggest, but not by much. There is considerable scope for estimates of resources to differ, by the very nature of the definition of a resource. The United Nations Intergovernmental Panel on Climate Change, the biggest body of international scientists ever to have studied these issues, calculates that coal significantly dominates the resource tally. By their reckoning, total coal resources exceed 3,000 billion tonnes of oil-equivalent, total gas resources exceed 2,500 billion tonnes of oil-equivalent (of which 55% is recoverable gas hydrates) and total oil resources exceed 1,000 billion tonnes of oil-equivalent (of which more than 80% is recoverable unconventional oil, mostly tar sands).[3]

The climate crunch

Incoming solar radiation is of shorter wavelengths than the radiation 'bounced back' out to space by the surface of our planet. Greenhouse gases in the atmosphere, such as carbon dioxide, let the incoming radiation through, but trap the outgoing radiation, thereby increasing the temperature.

This phenomenon, known as the greenhouse effect, is essential to our existence. With no greenhouse gases at all, the temperature at the surface of the Earth would be an inhospitable 33 degrees Celsius lower than it is today. But with the advent of fossil-fuel burning, carbon dioxide concentrations have built up steadily, along with the concentrations of other greenhouse gases such as methane (emitted by fossil-fuel use as well as agriculture and rotting waste). As the amount of carbon dioxide has built up from the pre-industrial concentration of around 280 parts per million to more than 386 parts per million today, so the global average temperature has followed an irregular and troubling upward trajectory.

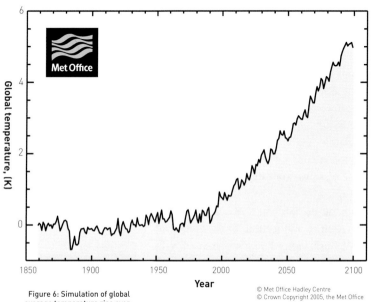

© Met Office Hadley Centre
© Crown Copyright 2005, the Met Office

Figure 6: Simulation of global average temperature rise over the 20th and 21st Centuries relative to the end of the 19th Century, using the Met Office Hadley Centre climate model including feedbacks from climate change on the carbon cycle. The model produces a realistic simulation of the 20th Century, but these feedbacks accelerate the atmospheric CO_2 rise into the future compared to the more usual climate simulations which neglect these feedbacks. Whilst the main set of climate model simulations used in the IPCC Fourth Assessment Report (IPCC 2007) neglected climate-carbon cycle feedbacks, the IPCC report acknowledged the potential importance of these feedbacks and discussed their uncertainties. Some other climate models also now include these feedbacks; the Met Office model shows the strongest feedbacks. British Crown Copyright (2009), reproduced by permission of the Met Office.

This process has been known about since 1827, and scientists have warned on occasion since then about the implications. But it wasn't until the late 1980s that concern really crystallised. It is instructive to consider the history of the global-warming issue from that time through to the present.

In 1989, the United Nations convened a panel of 300-plus of the best climate scientists from around the world to study the problem: the Intergovernmental Panel on Climate Change (IPCC) scientific working group. The first IPCC scientific assessment report was released in May 1990. At the press conference for the launch, then prime minister Margaret

Thatcher warned that the conclusions would 'change our way of life': that unless we acted we would cry out in the future not for oil, but water. At the time, the world seemed to be listening. The United Nations called for multilateral negotiation of a climate treaty cutting greenhouse-gas emissions, beginning with a World Climate Conference. Most governments signed up. But those talks have run for eighteen years now, and greenhouse-gas emissions have carried on rising.

The second IPCC scientific assessment, in 1995, narrowed the uncertainties in the first report, and added urgency to the scientists' warning. The IPCC's scientists are required by their governments to operate by consensus when writing their reports, and – notwithstanding this considerable hurdle – by 1995 they felt able to announce that they could see the first faint imprint of human enhancement of the greenhouse effect in the pattern of rising temperatures around the globe, something they did not feel able to conclude in the first report. This increased confidence allowed the negotiation of the Kyoto Protocol in 1997. The Protocol committed industrialised governments only to token emissions reductions, summing in principle to around 5% of total emissions from developed nations by 2012. But at least it was a start.

The third IPCC scientific assessment, in 2001, came at a time of great frustration. The Kyoto Protocol required ratification by a significant majority of governments. George W. Bush Junior had come into office in January 2001, and in March 2001 he not only refused to ratify the Protocol, but also pulled America out of formal participation in the ongoing climate negotiations. The third IPCC scientific assessment narrowed the uncertainties around projections of global warming still further, and upped the alarm level to such an extent that it persuaded the rest of the world to keep the Kyoto negotiating process alive even without the US at the table. As a result, the national ratifications continued and eventually, in February 2005, the Protocol came into force.

With the publication of the fourth warning by the IPCC scientific group, in February 2007, the continuing levels of denial in the Bush administration and its narrowing band of corporate supporters became ever more difficult to understand. The scientists 'most likely' estimate for the increase in global average temperature, in the absence of meaningful cuts in emissions, stood at 4 degrees Celsius by 2100, with the possibility of rises of up to 6.4 degrees Celsius once potential natural amplifers to the man-made warming – so-called positive feedback effects – are included. This is higher than the 2001 report's estimate, which was 5.8 degrees Celsius at the upper end of the range. Around 2,500 scientists in all collaborated on this fourth assessment. Six years before, in the third assessment, the scientists concluded that observed warming was 'likely' to be man-made. In 2007 this connection was deemed 'very likely'. It would have been deemed 'virtually certain' but for the insistence of China and a few others that the confidence level be watered down.

The writing on the wall ought to have been clear enough back in 1990. In December of that year, at the World Climate Conference – the UN event called to kick-start negotiations for a global climate treaty – environmentalists urged governments to go about their task as though it were a military security-threat assessment. In such an exercise, green groups argued, governments should base their policy response not on the best-guess consensus of the world's best experts, but on their worst-case analysis. They tabled a scenario wherein human greenhouse-gas emissions stimulated huge feedback emissions in nature, for example, from melting permafrost and drying soils and forests, none of which were incorporated in the climate computer simulations of the day. In the very worst case, the amplifications could lead to a runaway effect, where feedbacks drowned the potential to cut human emissions from fossil-fuel burning and other sources. Society needed to take out massive insurance against this horrific prospect, the green groups argued. Billions needed to be invested in renewable and energy-efficient technologies to replace fossil fuels, just as billions had been invested – rightly or wrongly – in taking out military insurance against a worst-case scenario of invasion by the Red Army during the Cold War.

Diplomats have gathered for eighteen years at the climate talks, trying to cut the emissions from fossil fuels. They have failed so far even to freeze them. But history is not destiny. Could peak oil, and the emerging solar revolution, be the catalysts for eventual multilateral success?
(Image: Jan Golinski, UNFCCC)

The entreaty was dismissed as scaremongering at the time, including by many of the IPCC's own scientists. But today, checking the feedbacks in that eighteen-year-old scenario against emerging collective understanding of climate science, as captured by the fourth IPCC scientific assessment, almost every box has to be ticked. Today, as a result of our near twenty-year delay in responding, we face at minimum a wide range of potentially catastrophic climatic changes – from rising sea-levels, which could engulf many of the world's cities and lowlands, and an increase in the intensity of both droughts and floods to a reduction in agricultural capacity. We also see worrying evidence that feedbacks not included in forecasts, such as methane emissions from permafrost, are kicking in. To delay ambitious emissions cuts any further would be to invite the unthinkable.

The oil crunch

Greenhouse-gas concentrations have built up in the atmosphere so quickly in large part because modern economies have become oil dependent from root to branch. It's not hard to see why. Oil is very energy dense: the energy locked into one barrel (around a seventh of a tonne) is equivalent to that expended by five labourers working twelve-hour days non-stop for a year.

This energy density, combined with its liquid form, has made oil pivotal to almost every process in modern society. The agricultural sector, for example, depends on oil across the entire value chain, from the field to the cellophane-wrapped product in the shops. To raise a single beef cow in the United States – from conception to plate – requires the direct and indirect use of around six barrels of oil.[4]

Yet despite the risks of such dependence, corporate and ministerial plans in most countries have long been based on the assumption that supplies of oil will continue to grow, continue to meet rising demand, and do so at generally affordable prices. In recent years that rather crucial premise has come into question. In this section, we briefly examine why.

The peak-oil problem

Peak oil is the point where further expansion of global oil production becomes impossible because new production coming onstream is cancelled out by production declines elsewhere. Beyond this point, the world will face shrinking supplies of increasingly expensive oil. In the conventional argument, favoured by OPEC, BP, ExxonMobil and many oil-industry institutions, the peak of production will not occur for around a quarter of a century or more. The decline will then begin, but it won't prove problematic because we will have had plenty of time to bring renewable energy and alternative fuels into play. In contrast, the early peak view, held by a growing minority, is that the oil industry has lapsed into a culture of overexuberance about both the remaining oil reserves and prospects of oil yet to be discovered, and about the industry's ability to deliver capacity to the market even if enough resources exist. The risk, this group argue, is that the world is sleepwalking into an unprecedented energy crisis, in which oil, so central to every part of the modern economy, fairly rapidly becomes either impossibly expensive, very difficult to obtain, or both.

The early peakers are led by a group of ex-oil industry geologists in the Association for the Study of Peak Oil (ASPO), members of which include many retired senior industry figures. In the corporate world they are led by the recently formed UK Industry Taskforce on Peak Oil and Energy Security (ITPOES), members of which include Virgin, Arup, Foster and Partners, Scottish and Southern Energy, Solar Century, Stagecoach and Yahoo.

The first report of the ITPOES group of companies, released in November 2008 at the London Stock Exchange, presented evidence that total global oil production will begin declining at some point in the period 2011–13. The main argument is that new capacity flows coming onstream from discoveries made (and publicised) by the oil industry over the preceding decade will begin to drop. This problem will be compounded by other issues, including the accelerating depletion of many of the old oilfields that prop up much of global oil production today, the probable exaggeration by OPEC countries of their reserves, and the failure of the 'price-mechanism' assumption that higher prices will lead to increased exploration and expanding discoveries.

In the same report, Shell presented a different view to the ITPOES companies. (Shell was not a member of the taskforce, but agreed to contribute its perspective on peak oil, in the same report, to help clarify the debate.) The Shell view is that the era of 'easy' oil will be over around 2015, and global production can be maintained on a plateau extending into the 2020s by resorting to unconventional oil resources, especially the vast deposits of solid tar in Canada known as the Athabasca Tar Sands.

Because there is widespread expectation of rising supply for some decades to come, these contrasting views would both present huge challenges for the world economy.

Concerns about conventional oil

In the same week the ITPOES report was released, the International Energy Agency (IEA) published its latest weighty annual report, the 'World Energy Outlook'. In 2008, for the first time, the IEA conducted an oilfield-by-oilfield study of the world's existing oil reserves. (One might reasonably ask why they had not done so before.) It revealed that the fields currently in production are running out alarmingly fast, as Figure 7 shows. The average depletion rate of 580 of the world's largest fields, all past their peak of production, is fully 6.7% per annum. Without investment in enhanced oil recovery (the various techniques petroleum engineers have of boosting recovery factors in their oilfields), the figure is 9%. Crude oil production from all the world's existing fields climbs unevenly from just below 60 million barrels a day in 1990 to a peak – more exactly a brief plateau – of just over 70 million barrels a day (9.5 mt a day) between 2005 and 2008. In 2009, however, crude production begins a steep descent,

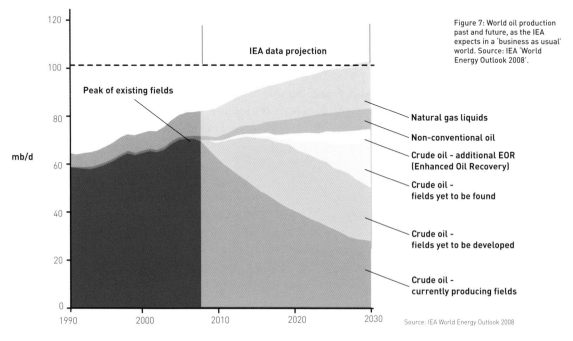

Figure 7: World oil production past and future, as the IEA expects in a 'business as usual' world. Source: IEA 'World Energy Outlook 2008'.

Source: IEA World Energy Outlook 2008

falling steadily all the way below 30 million barrels a day by 2030. The depletion factor might better be called a fast-emptying factor.

This is doubly alarming because, even with demand for oil being destroyed fast by recession in the West, the IEA expects the rate of demand growth – led by China and India – to be so high that the world will need to be producing at least 103 million barrels a day by 2030.

Can that be done? Only, says the IEA, if massive investment is thrown at the challenge of making up the shortfall, especially by the OPEC nations. Global production at the time of writing totals 82.3 million barrels a day if we subtract biofuels and add to existing crude production the 1.6 million barrels a day of 'unconventional' oil squeezed from the tar sands and the 10.5 million produced during gas-field operations. To reach production of 103 million barrels a day, therefore, would require oil-from-gas (oil that is produced as a by-product of natural-gas operations) to expand almost to 20 million barrels a day, unconventional production to expand to almost 9 million barrels a day, and on top of that an addition of 45 million barrels a day of crude oil capacity yet to be developed and yet to be found. All this adds up to 64 million barrels a day of totally new production capacity needed onstream within twenty-two years. That, the IEA points out, is fully six times the production of Saudi Arabia today.

At oil prices below around $70 a barrel, producing oil today becomes uneconomic in many settings. The oil price languishes at the time of writing at less than $50 a barrel. Pricing in the oil market has become completely disconnected from 'fundamentals' by the volume of 'paper trading', and oil development and exploration projects are currently being cancelled around the world on a regular basis. How on earth is the industry going to bring onstream six new Saudi Arabias from this kind of start?

Let us précis the additional concerns of the ITPOES group, and then consider the implications.

The oil industry is not discovering giant oilfields at anything like the rate it did in the 1960s, the peak decade for discoveries. This is the case even with much better equipment for exploration today, and even after four years of rising oil prices from 2004 into 2008, when exploration was not hampered by lack of funds for investment. When the oil companies do make big discoveries, the lead times from discovery to first new oil delivered to market are long: often more than ten years. The biggest discovery this century, the Kashagan field found in 2000 in the Caspian

Sea, was expected at the time to produce its first oil by 2005. Today, after endless delays, it is not due to come onstream until 2013. As mentioned above, the net global flow-rate projections from the oil industry's own public data slow down in 2013 unless global demand falls considerably (Figure 8). It is difficult to understand why this alone is not sounding alarm bells in governments and industry, much less when combined with the constant record of delays in bringing oil onstream.

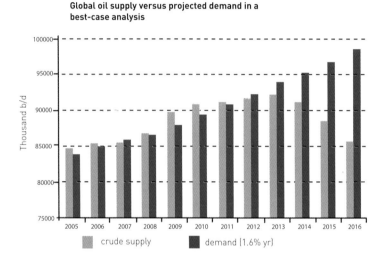

Global oil supply versus projected demand in a best-case analysis

Figure 8: Global oil supply versus projected demand in a best-case analysis, according to the UK Industry Taskforce on Peak Oil and Energy Security in its 2008 report. (see.www. peakoiltaskforce.org for more details.)

In addition, the oil industry has profound infrastructure problems, and major issues with underskilling and underinvestment. Many drilling rigs, pipelines, tankers and refineries were built more than thirty years ago, and according to some insider experts the physical state of the global oil infrastructure is a major problem even at current rates of oil production, much less the significantly higher levels anticipated in future. The average age of personnel in the oil industry is 49, with an average retirement age of 55: little less than a human-resources time bomb. To add to the challenges, the industry's overall exploration budget has actually fallen in real terms in recent years. The ITPOES fear is that these issues will synergise to compound the peak-oil crisis, gravely impairing society's collective ability to respond.[5]

When the full gravity of the oil crunch dawns on governments, there is scope for the peak-oil threat to relegate the climate threat, in the eyes of some policymakers. There will surely be further calls for expansion of production in the tar sands, and for major efforts to extract coal in other unconventional ways, ignoring the carbon implications. Others have advanced nuclear energy as a solution, thinking that, in an age of electric vehicles, the electricity can be nuclear. To see if those suggestions stack up, let us briefly consider tar sands, coal and nuclear in turn, and in the context of the climate crunch.

Conflating the climate and oil crunches

At first pass, worrying about both the climate-change crisis and the oil-depletion problem seems contradictory. The reaction of many people is to ask: if oil is depleting fast, why worry about greenhouse-gas production from the burning of it? But we should be concerned about both problems at the same time. According to the best advice available to governments, we tip the climate into chaos of the kind that can cripple economies and trigger potential points of no return beyond a 2 degree Celsius hike of the average global temperature. That maximum acceptable level of warming has been accepted as a target by the governments of the EU, among others. We have already turned the global thermostat up by 0.8 degrees from the pre-industrial average, so we have a maximum of 1.2 degrees to go.

To have a decent chance of staying below 2 degrees, we can 'afford' to emit no more than 400 billion tonnes of carbon, and probably much less. We have enough fossil fuels available in accessible resources to exceed the danger threshold by at least ten times.[6]

Let us assume that those who worry about an early peak in oil production are correct, and that there are far fewer resources and reserves of oil available than commonly assumed. Let us also assume that there is far less gas and coal left than assumed. We still have more than enough fossil fuel to trigger climatic ruin, especially in coal.

Tar sands: small flows, much carbon

There are vast amounts of oil locked up in the tar sands, and certainly hundreds of billions of barrels of it are accessible in principle. But the oil is difficult to extract. It is solid, not liquid, and has to be melted, mostly underground. That requires significant quantities of gas and water. Even then, progress is glacial. The oil industry has invested $25 billion to date,

and after decades of effort has a production capacity of 1.3–1.4 million barrels a day at present. Industry estimates now put production in 2015 at little more than 2.5 million barrels a day. It is difficult to see how that can make much difference if, as even ExxonMobil concedes, the annual depletion rate of easy-oil is already around 3.5–3.9 million barrels a day (which equates to 4–4.5% of the global total). The IEA's 'World Energy Outlook 2008' warns that depletion will accelerate. By 2015, what will the depletion rate be in conventional oil?

Well over $100 billion of new investment would be needed to ramp up the tar-sands production to the levels the industry foresees in 2015. In the face of these challenges, at least one oil company, Talsiman, has lost faith in the tar-sands proposition and pulled out.

The oil shales of Wyoming and Colorado are also held up by some as a considerable hope for the future. In this type of unconventional oil, organic matter has yet to be cooked to the level where it forms either crude or tar. As in the case of the tar sands, there is plenty of 'oil' there in oil shales, if it can be cooked underground. But how to cook it? Whether there is any realistic technique for doing so, or if so on what timescale, remain open questions.

Extraction of tar sands in Alberta, Canada.

Coal: far from capture and storage

Oil can be extracted from coal by pulverising the coal and passing gases across it at high temperatures, a process commonly referred to as coal-to-liquids. This is such an energy-intensive process that few have tried to commercialise it with any seriousness of intent, until recently. Coal-to-liquids plants are under way or planned in the US, China, Australia, India, Japan, New Zealand, Indonesia, Botswana and the Philippines.

Converting and burning the liquid from coal emits twice the greenhouse gas of burning diesel, meaning that there is a considerable environmental toll from the coal-to-liquids approach. In June 2007, China reportedly considered halting coal-to-oil projects due to worries about energy, expense and water requirements. The official Xinhua News Agency reported an official of the country's top economic planning agency, the National Development and Reform Commission, as saying that China 'may put an end to projects which are designed to produce petroleum by liquefying coal'. In August 2008 the country did just that: the Chinese government ordered a halt in all coal-to-liquids plants bar one in order to conserve coal supplies for power generation.[7]

Faced with this evidence of environmental and resource constraints, and only small flow-rates projected far in the future, it is difficult – as with tar sands – to imagine that coal-to-liquids can contribute significantly to closing the easy-oil depletion gap, even if environmental considerations are ignored. And such constraints, of course, will not be ignored.

Many favour using electricity from coal to replace oil, by charging electric vehicles, arguing that a technology called carbon capture and storage is a potential solution for carbon dioxide emissions from coal power plants. In this process, carbon dioxide is extracted from the coal either before or after combustion, and sequestered underground in a storage facility such as an abandoned oilfield or a saline aquifer. But there is a substantial timing problem. As US Energy Secretary James Schlesinger has put it: 'It will take fifteen to twenty years to introduce carbon capture and storage, if then.' The languor with which policymakers set their carbon-capture-and-storage goals would seem to support such a lengthy timeframe. Proposed EU legislation envisages all coal-fired power stations built in the EU having carbon capture and storage after 2020. EU leaders expect to commit to twelve large-scale pilot carbon-capture-and-storage projects by 2015.

There is also the question of whether or not carbon capture and storage will work, even if it proves ultimately deployable at industrial scale for the 2,000-plus power plants that may be built or revamped by 2020 on current trends. In the UK, the government appears likely soon to license the first British coal plant to be built for thirty years, at Kingsnorth in Kent, provided it is made 'capture-and-storage-ready'. However, tellingly, the Department of Business will not require the plant operator, E.ON, to fit capture-and-storage equipment by a target date. The reason given for this is that E.ON wouldn't go ahead with the plant if the government set a cut-off date for capture-and-storage operation, given that they are not certain that the technology is going to work, that the engineering has not been tested and no-one is fully aware of what the costs might be.[8] This approach speaks volumes for the practicability of the capture-and-storage option. It suggests – at minimum – that risk-abatement concerning coal should not be focused on an aspiration of storing carbon dioxide underground, and certainly not at the expense of market-ready energy solutions. That said, capture and storage should be quickly and thoroughly investigated for potential use. In fighting climate change, we will need every weapon available.

Except perhaps one.

Nuclear power

Some people argue that nuclear power can play a key role in combating climate change, and that oil depletion means we will need nuclear to provide electricity for electric vehicles to replace those powered by the internal combustion engine. The problems with this argument include timing and the diversion of finance from energy technologies that can provide answers more quickly and safely than nuclear can.

During 2008, captains of the nuclear industry acknowledged that they cannot build and bring onstream the next generation of power plants in less than ten years. That simply isn't fast enough to make a difference either to our easy-oil depletion problem or our clear and present climate-change danger. In 2018, new plants would be coming onstream seven years after the most likely year that the peak-oil crisis dawns on the world. Then, they would be replacing a bare minority of the 429 nuclear reactors active in the world today, many of which are already near or past their supposed decommissioning dates as things stand.

The nuclear industry is also quite probably being over-optimistic – and not for the first time in its half-century history. The first European nuclear plant to be given a go-ahead in ten years was the Finnish Olkiluoto 3 plant, in 2002. This is a Generation III plus European Pressurised Water Reactor (EPR), a modified and larger (in terms of output) version of the Generation III pressurised water reactors that are common throughout France. The Finnish plant had a projected completion date of 2009 on a budget of €2.5 billion. Completion is now estimated not sooner than 2012, and the station is €2.2 billion over budget. This could increase as the operator TVO seeks compensation against Areva, the vendor, for the delays. Another Generation III plus EPR reactor, at Flammanville in France – the first reactor to be built in the country for fifteen years – is also in trouble. It is currently 20% over budget. Meanwhile, advocates of further French nuclear expansion await the results of a French government programme to test the groundwater under all fifty-eight French nuclear reactors, after a recent spate of radiation leaks. Such setbacks may not be surprising. The nuclear industry has a major skills shortfall because of the long period it has spent in enforced abeyance since Three Mile Island and Chernobyl.

Even if it could make a difference in the coming global energy crisis, the industry hasn't yet found a way to deal with its radioactive wastes, after half a century of trying. Efforts by government to help them get round this inconvenient truth require hidden subsidies that would effectively be blank cheques extending far into the future. But to commit to such subsidies would be to take vital resources from the genuine short-lead-time technologies. There is only so much capital that can go to energy, given all the other calls on governmental, corporate and household budgets, and within the energy budgets only so much can go to each energy technology. This is especially so in a recession. There should be no mistaking the choices involved here. Channelling billions of dollars preferentially to nuclear removes billions from energy conservation, energy efficiency and renewables, including solar. This unstated competition is why many nuclear supporters have tried for so long, in the editor's belief and experience, to play down and generally suppress renewable energy.

All told, then, neither nuclear, tar sands nor carbon capture can be relied upon to solve the twin threats of climate change and peak oil. The

best we can hope for is that enough oil and gas continues to flow in the coming years to allow the world economy to keep functioning. That will provide the global community with the window of opportunity it needs to embark upon a massive-scale rollout of clean, sustainable energy sources – nothing less than an energy revolution.

The rest of the book considers this energy revolution, and in particular the role of solar energy.

Sizewell nuclear power station, UK.

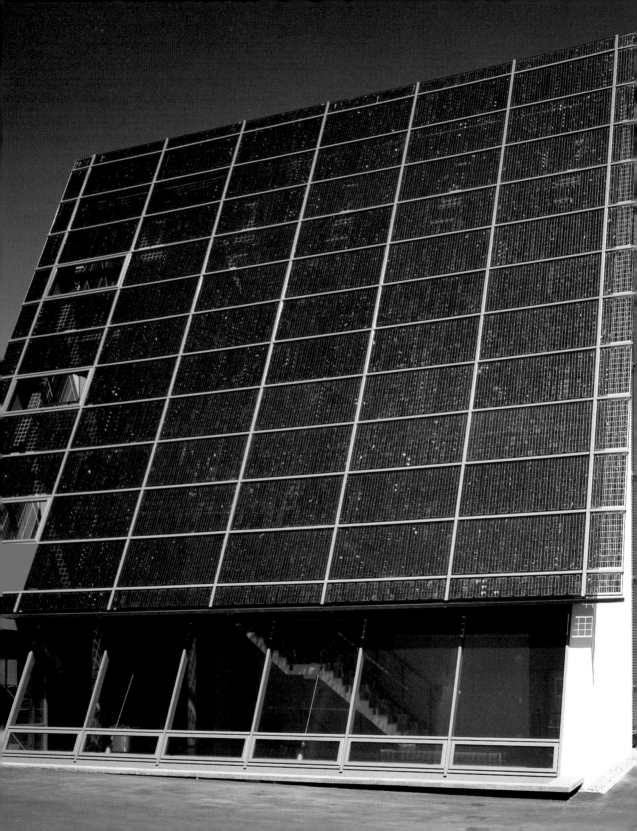

Chapter 3

THE ENERGY REVOLUTION

How Solarisation Can Power the World

In this chapter, we consider the global scope of renewables, the role of the electricity grid in the renewables story, and key innovations that amplify the effectiveness of renewables: the smart grid and improved renewable-energy storage. The aim of the chapter is to paint a picture of what renewables are capable of, as context for the detailed discussion of solar technologies that follows in the next chapter.

Capability is pointless without desire for change, and we end the chapter with a discussion of public opinion on renewables and the way the world uses energy.

Solarisation

In 2001, planners from a major corporation calculated the available energy resource for the main renewable-energy technologies in all the main regions of the world. They found that renewables hold the potential to power a future world populated with 10 billion people. This could be achieved even if all of them used energy at levels well above the average per capita consumption today in the EU – something that should never be necessary given the savings the world could make by increasing energy efficiency. Simply stated, renewables could easily get us to zero carbon emissions while making energy insecurity a thing of the past. What made this important study all the more noteworthy was that the corporation in question was Shell, one of the world's biggest oil companies.[1]

Shell's calculations showed that, in terms of resource, solar power comes first as a potential contributor to a renewables-powered future, followed by geothermal energy, which involves extracting energy from

Opposite, Solar building facade in Kassel, Germany.
(Image: SMA Solar Technology AG)

hot rocks. It should not come as a surprise that solar tops the list, when you consider the arguments set out in Chapter 1.

The geothermal resource is large partly because buried pipes can be used to transfer heat from the ground into buildings, and not just in volcanic areas. Geothermal heat pumps can also be used in reverse in hot seasons, to pump heat out of the building into the ground, thereby obviating the need for additional air-conditioning equipment and also returning heat for use in the next cold season.

Wind power can also play a major role. Wind-power stations, often called wind farms, exploit the rotation of aerodynamic blades to drive turbines that create electricity. The US could in principle provide all the electricity it uses today from the wind-power potential of just three states: Texas, North Dakota and Kansas. Europe's demand could be met using offshore wind farms alone. In the UK's case, only a tiny fraction of suitable offshore areas would be needed to meet the nation's current demand.

As already mentioned, most of the renewable technologies are solar technologies in the broad sense: they are driven either directly or indirectly by sunlight falling on Earth. Solar, wind, hydro, wave and biomass are called 'renewable' for a good reason: they hold the potential for generating power as long as light falls on the planet.

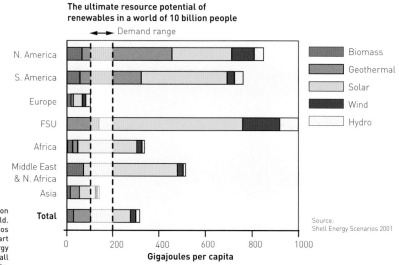

Figure 9: How solarisation could run the world. Taken from the Shell scenarios published in 2001, this chart shows the theoretical energy production by renewables for all the main continents.

Source:
Shell Energy Scenarios 2001

Options with and without electric grids

Most electricity in the world moves from the places it is created to the places it is used via grids. A grid is an interconnected web, usually with generators at one end and customers at the other. In between is a transmission system of high-voltage wires and a distribution system, into which the transmission system feeds via substations that reduce the voltage before electricity is delivered to users along the final part of the grid, where consumption is measured using meters.

Most electricity today is generated in giant central power plants burning fossil fuels, or using nuclear materials. But one very useful thing about solar photovoltaics and solar thermal, when used for electricity generation, is that they can be deployed along the transmission system and can feed directly into it. So too can other 'micropower' technologies such as small combined-heat-and-power plants powered by biogas or natural gas, fuel cells and other types of renewable-power technology. Technologies that feed directly into the transmission system this way are known as 'distributed generation' (also known as embedded generation). The more conventional model, where large fossil-fuel power plants are used as generators, is called central generation. Whereas a central power plant might be in the 500 megawatt to 3 gigawatt range, a distributed generator may be as small as a kilowatt (solar panels on the roof of a small house, say), ranging up to a few megawatts.

Distributed generation has a number of advantages over central generation. It reduces the need for transmission and distribution lines and their upgrades. It avoids the inherent transmission losses associated with large grids – potentially up to 30%. It provides energy security by diversifying the types of energy used. It increases the reliability of the grid. And it lends itself to a modular approach, so new increments can be added as a user's requirements grow.

If we don't connect renewable technologies to the grid, and use them instead to create electricity that can be stored in some way (see p.38 for options), we have created what is known as an off-grid system. Off-grid systems are useful wherever extension of the grid is problematic. In much of the developing world, extension of grids is unaffordable even where practicable. In such places, off-grid solar systems are the best option, for communities and households alike. Batteries are the most common form

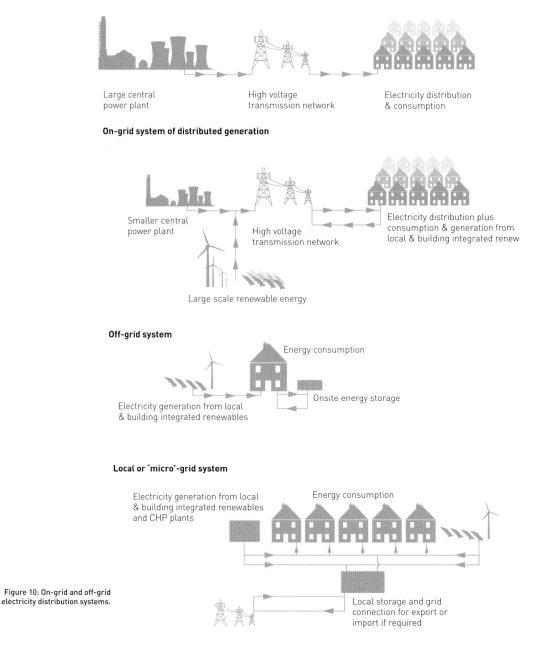

Traditional on-grid system of centralised generation

Large central
power plant

High voltage
transmission network

Electricity distribution
& consumption

On-grid system of distributed generation

Smaller central
power plant

High voltage
transmission network

Electricity distribution plus
consumption & generation from
local & building integrated renew

Large scale renewable energy

Off-grid system

Energy consumption

Electricity generation from local
& building integrated renewables

Onsite energy storage

Local or 'micro'-grid system

Electricity generation from local
& building integrated renewables
and CHP plants

Energy consumption

Local storage and grid
connection for export or
import if required

Figure 10: On-grid and off-grid
electricity distribution systems.

of energy storage in these kinds of installations. With the right channels of distribution and credit in place in the developing world, off-grid solar home systems are affordable at lower monthly outlay than households would normal spend on the alternatives such as candles and kerosene.

A final type of grid falls somewhere between a centralised national or regional power grid and an off-grid system. This is a local grid with enough distributed generation on it to keep electricity supply going at all times. This arrangement is referred to as a local- or micro-grid, or a private-wire network. Existing private networks, like those in the UK town of Woking, have proved completely reliable. However, they can be connected via a spur to the national grid – just in case.

The ability either to link small solar generators on grids, or to operate them off-grid – storing solar electricity for later use – greatly amplifies the 'power of light'. We explore this further in Chapter 5, when we look in more detail at solar in buildings, and their capacity to become stand-alone power stations.

The use of micropower is growing fast around the world. A third of the world's new electricity capacity in 2006 was provided by waste-heat or gas-fired cogeneration, wind and solar power, geothermal, small hydro, and waste- or biomass-fuelled plants. At least a sixth of the world's total electricity now comes from micropower rather than from central thermal stations. In twelve industrial countries, micropower now provides from one-sixth to over half of all electricity.[2]

The history of electric grid systems

If distributed generation has so many advantages, why has the world ended up after more than a century of electricity generation with centralised power systems dominating almost every country? The answer lies in the outcome of a struggle more than a hundred years ago between rival electrical technologies, alternating current (AC), in which the direction of the electric flow continually switches back and forward, and direct current (DC), in which the current flows in one direction only.

Thomas Edison, the great late-nineteenth-century inventor, envisioned a world where electricity would be generated by many small micropower plants generating DC electricity that could be used directly in homes and workplaces. The plants would be controlled and run locally, and would deliver power to the people both literally and metaphorically. Another great inventor, Nikola Tesla, believed at the same time in a system run on AC. High-voltage AC power could be transported more efficiently than DC in wires over long distances from large power plants.

Thomas Edison.
(Image: L.C. Handy 1877)

Edison's DC model was backed by interests including banking colossus J.P. Morgan, while Tesla was supported by Pennsylvanian businessman George Westinghouse. By the dawn of the twentieth century, AC had won an outright victory. Edison's company, Edison Electric, morphed into modern giant General Electric, which went on to build its empire on AC infrastructure. Westinghouse's company today makes nuclear and other power plants for pumping AC into high-voltage transmission systems.

The early builders of the AC empire bought up the small electricity companies that could have delivered Edison's vision and argued that a national electricity system should be a monopoly (though it should be noted that AC can be used with equal facility on electricity grids dominated by either centralised or distributed power plants). Politicians by and large went with them, concentrating power in the hands of relatively few people. At the beginning of the 1930s, for example, just eight companies controlled nearly three-quarters of America's electricity supply. This led to inevitable excesses, and forced government to break companies up and set up regulatory bodies. The outcome has been decades in which those who have sought to exploit the tendency to monopoly in central generation have battled with regulators who have sought to constrain it.

Nikola Tesla.
(Image: Library of Congress)

Compared to the simplicity of Edison's vision, and the extent to which it would have engendered local empowerment, the global electricity system has, in the view of some, been dysfunctional for a long time. Whether that is true or not, technology has evolved to such an extent since the days of Edison and Tesla that major new opportunities have arisen in the Edison model. In particular, as we will see in the next two sections, storage innovations and so-called 'smart grids' allow electricity systems to be operated much like the internet, cutting out many of the arguments for centralised power plants and favouring distributed generation.

Intermittency and storage

Distributed renewable power generation in the grid is not challenge-free. The frequency of the grid's alternating current needs to be set at a certain level in order for it to operate effectively. Renewable generators tend to fluctuate in output, for example, as the wind varies in speed, or as cloud cover comes and goes, blocking sunlight. As the proportion of distributed renewable generation rises, this intermittency can cause the frequency to fluctuate outside the operating range. On top of this, energy demand is not constant: it varies throughout the day and the year, and there can be peaks of extreme demand. An energy provider needs to make sure there is a big enough supply of energy to cover these peaks, otherwise there will be blackouts. Building sufficient central generating capacity to cover the peaks is expensive and inefficient, as the peak capacity is required only rarely, meaning some power plants stand idle for much of the time (or as 'spinning reserve', with their turbines running like a car ticking over in neutral). A much better solution is to have some stored energy on hand to cover the peaks. Stored energy can also help deal with the frequency fluctuations arising from intermittency.

Given all this, it is not surprising that many ingenious ways have been developed to store energy. Let us quickly review some of them. Our intention here is not to be comprehensive – there is much innovation under way in this area – but to give newcomers to the energy debate a sense of the range of options, and the ultimate potential.

Batteries

A battery is a device that can convert electrical energy into chemical energy and back again. Charging batteries is useful at all scales from consumer electronics and off-grid solar roofs to back-up power for utilities. At the time of energy demand a chemical reaction is instigated in a substance within the battery called an electrolyte, creating electrons. In lead-acid batteries – the most common type of battery in use today for home-energy storage – the electrolyte is a liquid (dilute sulphuric acid). Such batteries tend to be bulky and small, and are best used in homes.

In a second type of battery, using vanadium as the electrolyte – a vanadium redox-flow battery, to use its full name – stores the electrolyte in external tanks, and circulates it through the battery stack as it is needed to generate energy. These batteries can be much bigger than lead-acid batteries – garage-sized and upwards, linkable in a modular way much like solar cells – and can provide load-levelling capacity on the electricity grid. Though they are still in early development, such batteries have long lifetimes, low operating costs, and are in use today. At a wind farm in Japan, a flow-battery system can supply 4 megawatts of power for up to 1.5 hours. One US electricity provider uses a 250-kilowatt battery to balance peaks for 8 hours uninterrupted. Manufacturers speak of storage costs as low as 10 US cents per kilowatt-hour for a family residence, once the technology is being manufactured at scale. If such prices are reached, solar photovoltaics on roofs will have become a nocturnal energy source.[3]

A third category of battery, advanced batteries, includes lithium-ion, nickel metal hydride, and sodium sulphur types. These use solid electrolytes, and take up less space than lead-acid batteries. At current levels of production they are too expensive for utility applications, but are used widely in consumer electronics and electric vehicles. As the use of electric vehicles grows, so the volume of production of advanced batteries will rise, and the cost will fall, potentially allowing their use in utility-scale projects as well.

Hybrid vehicles (those with petrol engines and batteries) and electric vehicles can also be used as storage devices. American clean-energy guru Amory Lovins has long held the idea that plug-in hybrid cars can be converted into mini-power plants on wheels, each of them with a generating capacity of 20 kilowatts or more. Batteries can be charged by day when their owners are working or otherwise occupied, by simply plugging them into a renewable source of electricity, and the cars can then connect to the grid to discharge any electricity not needed, providing their owners with useful income while buffering the grid.[4] The first plug-in hybrids come onto the market in Japan in 2009. Of course they still use gasoline and emit carbon, but in far smaller quantities than conventional vehicles.

Batteries are widely used today in buildings such as hospitals and data centres that need to ensure uninterrupted supply in the case of a grid power-failure. In case of outages, battery storage provides immediate power, which is then often replaced by diesel generators once they have had time to start up. These systems are known as uninterrupted power supplies.

Hydrogen and fuel cells

Hydrogen can be used as a chemical energy-storage medium. Energy is used to create pure hydrogen; when the energy is required, the hydrogen can be burned as a fuel or turned into electricity using a device called a fuel cell.

While hydrogen is the most abundant element in the universe, it is not found naturally on Earth except when combined with other elements in molecules such as water and methane. A range of technologies can be used to separate pure hydrogen from these molecules though today virtually all hydrogen is produced by one process: the steam reforming of methane, in which methane and hydrogen are combined to produce hydrogen and carbon monoxide.

In a future 'hydrogen economy' it is often envisaged that hydrogen fuel, produced primarily to replace transport fuels, would be created centrally in large nuclear- or fossil-fuel plants coupled with carbon capture, or locally on a smaller scale by electrolysis. The latter technique involves using two electrodes to pass electricity, preferably from low-carbon sources, through water, splitting it into oxygen and hydrogen.

As with all energy-conversion processes there are thermodynamic and other losses, and typically commercial electrolysers convert less than 70% of the electrical energy used into hydrogen. Further energy losses are involved in storing the hydrogen as pressurised or liquefied gas, or as a solid in a metal hydride.

A fuel cell is essentially a battery that recombines a fuel with oxygen on a catalytic membrane to produce an electric current. Fuel cells are one of the most efficient converters of chemical energy into electrical energy. If operating with hydrogen as fuel, they produce a single waste product: pure hot water. There are several different types of fuel cell, operating at different temperatures and with different fuels (such as hydrogen, alcohols, fossil hydrocarbons and even carbon) but they all work in this same basic way. They are available commercially for a variety of stationary, portable and transport applications where particular circumstances can justify their present high cost and relatively short lifetimes. Fuel cells for transport generally operate on hydrogen and utilise a small amount of platinum as the catalyst. Efficiencies vary, but typically about half (35–70%) of the chemical energy in the fuel can be realistically converted to electricity.

Hydrogen can be made from any electricity source, but it only makes sense in terms of carbon emissions if renewable sources of electricity are used. Even then, the energy losses in the fuel chain currently mean less than a quarter of the input energy does any useful work on the road. In contrast, electric vehicles using advanced batteries can leave more than two-thirds of the electrical energy for use on the road.

It is common for energy use to be inefficient. For example, for fundamental thermodynamic reasons, most coal-fired power plants only turn around 30–40% of the original heat energy from coal burning into electricity, which then is subject to further losses (around 7%) in the transmission and distribution system. The 'net energy' problem with fuel cells will need appreciable further innovation. There are also problems with the catalyst, which is commonly platinum: a valuable commodity in limited supply, currently extracted from only five mines in the world.[5]

Many billions have been invested globally trying to improve the performance and close the cost gap between fuel cells and their competition. Low production volumes and weak market drivers – rather than technology – are at the root of this challenge. Bulky fuel cells have already been installed in some buildings, generating both electricity and hot water, and small numbers of superbly engineered fuel-cell vehicles are on sale in various countries. But commercial progress with hydrogen, and hydrogen fuel cells, has disappointed many observers since the turn of the century, when expectations were very high. Increasingly opinion is centring on advanced batteries as the energy-storage technology of choice for the vehicles and off-grid solar buildings of the future. Caution must be advised, however, as battery technology, like fuel cells, has overpromised and underdelivered in the past, particularly in respect to energy-storage capacity. Many capable technologists are working on the hydrogen challenge, and the 'hydrogen economy' still has passionate advocates. This form of energy storage may yet prove to be more than a minor player in the solarised low-carbon future.

Other forms of energy storage

A popular and economic large-scale energy-storage solution is to pump water uphill to a reservoir in off-peak periods, and later to use the gravitational energy thereby stored up for hydroelectric generation. When demand rises, sluice gates open and extra power can be generated within seconds as water flows downhill through hydraulic turbines. This is the largest form of grid storage available at present, with almost 100 gigawatts of pumped water-storage capacity globally, or about one and a half times the total UK electricity-generating capacity.

Another kind of mechanical storage, much eyed for the future though only rarely in use today, involves using off-peak electricity to power a generator driving compressors that force air into an underground storage reservoir, such as a cavern in rock salt. As with pumped-water storage, the process is reversed at times of peak demand. A difference is that the air returning has to be heated using natural gas in order to power the generator, but much less gas is needed to produce a unit of electricity this way than is the case in a traditional gas turbine.

It is easier and more efficient to store energy as heat energy than as electrical energy. Buildings with solar hot-water systems, for example, can store heat from day to day using a very well-insulated water tank. A more ambitious approach is to store heat from one season to another. This is called a seasonal heat store, where the summer sun is used to heat a large insulated reservoir of water, using normal solar-thermal collectors. In the winter, the heat is used typically for space heating rather than hot water. Molten salt can also be used to store heat energy, as we will see in the section on solar-thermal power plants, on p.73.

One storage technology to watch is the flywheel, which is a cylinder spinning at very high speeds, storing progressively more mechanical energy the faster it spins. Energy can be drawn off by slowing the flywheel, or more energy can be stored by accelerating it. By using carbon-fibre rotors in a vacuum chamber, aerodynamic losses can be minimised, and with electromagnetic bearings losses from friction can be virtually eliminated.

Another storage technology for which there are great hopes is the supercapacitor. This is an electrochemical device somewhat like a battery. Whereas a battery undergoes a chemical reaction when charged, in a supercapacitor there is no chemical reaction, and the energy is stored as an electrostatic charge (or concentration of electrons) on the surface of a material.

We've touched briefly on just some of the energy-storage techniques that are already out there or in the pipeline. These will prove invaluable as solar and other renewables are rolled out more widely. Of course, the more efficiently and intelligently renewable energy can be used the less need there will be either for generation or storage capacity. This is what is particularly exciting about our next topic: the smart grid.

The role of the smart grid

It is often said of the traditional electricity grid, even in countries like the US and the UK, that it is the least impressive and most out of date part of the infrastructure. Politicians of all persuasions have recently joined with technology businesses to form an undeclared alliance intent on both upgrading it and making the way it works more intelligent. Advances in hardware and software during the digital revolution have made this process long overdue.

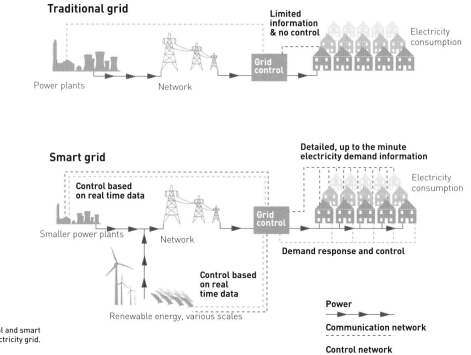

Figure 11: Traditional and smart electricity grid.

The 'smart grid' is best thought of as a megatrend within electricity markets, rather than a particular set of technologies. Today, the whole electricity-value chain from centralised power plant to building is subject to only selective central information-gathering and control. In a smart grid, operators will have a full grid perspective set out on their computer banks, all the way from the power plant to the many devices involved in end-use of electricity. They will be able to tune the whole system to minimise emissions and maximise efficiency. The operators' own profits will increasingly come from these services, and less from generation.

Let us consider some examples of where changes will come. Today, most meter reading is still done manually (if the meter is read at all: many bills are still estimated). In the smart grid, operators will enjoy automatic data collection, thanks to 'smart meters' in homes which will transmit information either wirelessly or via the electricity cables themselves. In today's grid, if a utility needs to reduce peak load, it needs to issue an appeal on local radio. In the smart grid, it will be able to tune load by turning end-use appliances off, or down, without inconvenience to the end-user. Turning off a compressor in an air-conditioning system for fifteen minutes, for example, won't significantly affect the temperature in the house.

Individual end-users will be similarly empowered: they will be able to use smart meters to adjust their microgeneration, for example, to load requirements in the building. Much of this will be automated. Built-in controllers in electrical devices like fridges will allow the device itself to respond to varying conditions in a grid. Should grid frequency drop, say, the fridge will be switched off by its controller as it detects the change – but not for so long as to allow food to spoil.

In electricity grids today, there is relatively little embedded generation and energy storage. In smart grids there will be plenty, all of it providing data allowing for central coordination. As electric vehicles proliferate, they will increasingly act as tuneable mini-power plants within the smart grid. Parked electric cars will be plugged into the grid, either charging or – if charged – being used as paid sources of electricity. In this way, intermittency will be countered, redundancy will be added, peaks will be further shaved.

The condition of the grid itself will be much better monitored than it is today. Local substations will be automated and controlled remotely. The ever diminishing number of large power plants at the top end of the value chain will be fully integrated with micro-grids to provide only what energy is needed, and – beyond a prudent reserve – no more.

These are just some examples of what will become commonplace. They show why many people think of the smart grid as the electricity equivalent of the internet. Indeed, the term 'intergrid' has entered the energy lexicon.

One can easily imagine the scope for value creation in this vision, as far as the consumer is concerned. The whole house will be programmable to maximise cost and carbon savings. The electric car will be chargeable at night, when rates are low. Household electrical devices will be programmable to turn off around times of peak demand, when utility prices are highest. Carbon savings will be monitorable all along.

This vision is set to become reality. Political leaders increasingly advocate it as a key route out of recession. In the US, President Obama intends to spend billions on the smart grid. In Europe, up to €200 billion is scheduled to be invested in transmission and distribution networks by 2020, and as things stand some €90 billion of that will be spent on smart-grid technology. In the US, an investment boom was under way in smart-grid technology at the time of the financial crash in September 2008, with a flurry of venture-capital deals in technologies for applying IT to power-grid monitoring and management. The prize is big. US regulators estimate that a 5% improvement in the efficiency of power grids would remove the need for forty-two large coal-fired power plants.[6] As efficiencies improve, and the need for traditional power-generation drops, so the goal of powering entire countries with renewable electricity and heat becomes ever more feasible.

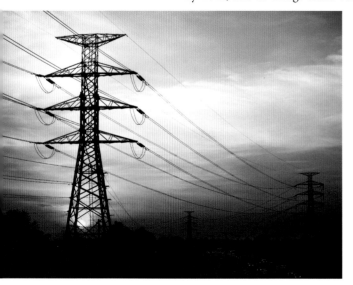

Could the sun be setting on traditional grid networks?

Powering entire nations

A number of recent studies have shown how entire nations can be run on renewable energy alone. Some have suggested, indeed, that nations could be run on solar energy alone, though this is not what the writers of this book advocate – we want to see the whole renewables family operating in strategic harness. Let us consider two such studies.

The first, by solar-thermal product designers with collective decades of experience of the technology, calculates what would be needed for existing solar-thermal technology to replace all US fossil-fuel electricity generation and petroleum-based transport, on its own, without any help from other solar or renewable-energy technologies. The study begins by asking what would be required to meet the 2006 US national grid generating capacity of just over 1,000 gigawatts, and peak-load capacity of 789 gigawatts, using a solar-thermal technology of the type recently announced by Californian utility Pacific Gas and Electric for a 177-megawatt project in California. This makes use of long mirrors known as heliostats to reflect the sun onto long steam-pipe receivers on towers (see the following chapter for more on this kind of concentrated solar technology). The calculations show that meeting 2006 US grid requirements would require covering 23,418 square kilometres, or a square 153 kilometres by 153 kilometres, with solar plants. To meet the requirements of an entirely electrified national land-transport sector, travelling 10 trillion miles as the US vehicle fleet did in 2006, would require increasing the square to 182–211 kilometres on one side.

Of course the solar-thermal plants would have to be in the sunny south-west, and power would have to be transmitted to the cooler north. This isn't as inefficient as it sounds at first impression, because winter home-heating loads in the north are mostly met by non-electrical generation (gas and oil). Air conditioning, however, needs a lot of electricity, but it's primarily used in the summer, when solar generation is at its maximum.

All this would cost between $4.5–6 trillion in capital investment, at today's prices. At an oil price of $100 a barrel – to take a round and probably conservative figure for the coming decades – the 13 million barrels a day the US imported in 2006 would cost almost half a trillion dollars a year. The 'payback' on avoided oil imports would be just 9–12 years, even before the massive coal and gas savings were factored in.[7]

There have been similar proposals in Europe, where some energy experts advocate connecting the European grid to huge new solar installations in the Sahara Desert. The land required to supply all the electricity or energy for the whole of Europe is of the same order of magnitude to that needed for the US.

An area 200 by 200 kilometres given over to solar thermal is far from unimaginable. But suppose PV were to be added to the equation. Or, for that matter, a mixed portfolio picked from the renewables family?

In January 2008, three long-serving solar researchers at government and other laboratories published a plan in *Scientific American* that, by 2050, would deploy 3,000 gigawatts of solar-photovoltaic farms in the American south-west, plus some large solar-thermal plants, with the daytime excess energy produced stored as compressed air in underground caverns for tapping at night to drive turbines. A direct-current network would carry solar electricity across the country, as in the first plan, delivering 69% of the US's electricity by 2050 and 35% of its total energy. Adding wind, biomass and geothermal energy could take the total to 100% of the electricity and 90% of the total energy, cutting greenhouse-gas emissions to the bone, and obviating foreign-oil imports early in the process.[8]

Two studies by researchers active in the solar-advocacy business do not necessarily make the case that nations can be entirely powered by renewables. But when economics ministries add their voice to the argument, even cynics might be tempted to listen. One of the loudest arguments of those who profess that traditional energy is needed even if renewables markets grow large is that modern nations cannot be powered properly without it. Renewables, they profess, cannot provide the minimum amount of power that a utility or distribution company must make available to its customers, otherwise known as baseload. But a German Economics Ministry-funded experiment showed in 2007 that distributed power can indeed produce baseload in a secure and reliable manner. Three companies and a university conceived and ran a 'Combined Renewable Energy Power Plant' experiment aiming to show in miniature what could be done if the will was summoned on the national scale to replace both fossil fuels and nuclear power.

The researchers linked thirty-six decentralised wind, solar, biogas, combined-heat-and-power and hydropower plants in a nationwide network controlled by a central computer. Using detailed weather data, they turned up the biogas and the hydropower, the latter in the form of pumped storage, whenever it was necessary to compensate for wind and solar intermittency. The system was scaled to meet 1/10,000th of the electricity demand in Germany, and was equivalent to a small town with around 12,000 households. It worked perfectly, meeting both continuous baseload and peakloads round the clock and regardless of weather conditions. The network was capable of generating 41 gigawatt-hours of electricity a year. Over the period of the experiment, 61% of the electricity came from eleven wind turbines (total 12.6 megawatts

Solar PV in Baden Würtemberg, Germany.
(Image: SMA Solar Technology AG)

capacity), 25% from four biogas CHP plants (total 4 megawatts capacity), and 14% from twenty solar PV installations (5.5 megawatts capacity). During the day of the press conference to announce the results, there was no wind at all in Germany and the country was covered by cloud, and still it worked. The results of the experiment suggest that, by 2020, 40% of German power demand could be met with wind, solar and bioenergy, and 100% by 2050. The current cost of generating electricity from the combined power plant is 13 euro cents per kilowatt-hour, twice as expensive as conventional electricity (which of course counts the carbon cost as zero, currently). But then the price of conventional polluting electricity is rising fast in Germany, as it is everywhere else, and carbon costs must surely be incorporated in the economics before too long.[9]

The Germans are not alone in making good progress towards a renewable-powered future. As mentioned in the Introduction, Portugal has a clean-tech plan to wean itself off oil. Its target, en route, is to generate 60% of electricity and 31% of primary energy from renewables by 2020, up from 20.5% of primary energy today, and the country is well on the way. Along the Spanish border in northern Portugal, where the world's biggest wind farm is under construction, turbine blades are built nearby in a factory employing over 1,000 people. The intention is to bring much more renewables manufacturing into the country in this way. A centrepiece of the solar part of the programme is the 45 megawatt Moura solar PV farm, a $500million project twice the size of London's Hyde Park wherein 2,520 panels, each the size of a house, track the sun through 240 degrees at a permanent consistent tilt of 45%. This project will be twice the size of any other solar farm in the world. There's action in the sea, too: three Pelamis 'sea-snake' wave generators near Porto began pumping electricity in late 2008. As a result of all this, Economics Minister Manuel Pinho dismisses nuclear power. 'When you have a programme like this there is no need for nuclear power. The relative price of renewables is now much lower, so the incentives are there to invest.'[10]

In the face of all this emerging evidence that economies can be run on renewables, attitudes are changing fast. *The Economist*, historically a conservative organ, is a good example. It recently offered the opinion that people who think we can't stop using oil suffer 'a failure of imagination'. An editorial in June 2008 waxed lyrical about the clean, green energy technologies that are so exciting Silicon Valley, and concluded that governments should tax carbon and drop subsidies for fossil fuels so as to accelerate the inevitable switch to these technologies around the world.[11]

When one understands the enormous scope that the smart grid and energy storage have to improve on these kinds of beginnings, it becomes easier to believe that a fundamental switch to solar and other renewables is possible in global energy markets.

As for the desire for change, that is certainly there, as we will see in the next section.

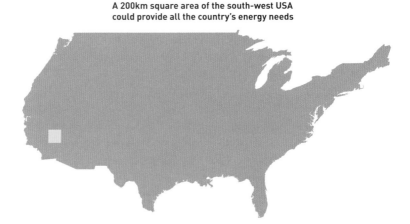

A 200km square area of the south-west USA could provide all the country's energy needs

Figure 12: The area of the US south-west needed in principle to generate enough solar-thermal energy to meet all US energy needs including transport.

Thirst for change

When people are asked about their favourite sources of energy, renewables consistently come out top, well ahead of fossil fuels and nuclear. Among the renewables, solar consistently comes out top. This popularity tends to span political divides. Polling in the US, for example, shows huge majorities of both Republicans and Democrats favour renewables spending. In 2006, 58% of Americans rated 'dealing with the nation's energy problem' a top priority, up from 40% in 2003. Around 80% favoured increased federal spending for research on wind, solar and hydrogen (with Republicans on 82% ahead of Democrats on 77%). This two-year-old poll was conducted when the oil price stood well below the record levels it subsequently rose to.

With the rapid rise of oil prices between 2004 and 2008, many people in many countries have come to think that oil is running out fast and major effort is needed on renewables as a consequence. In a recent survey of people in sixteen countries, majorities in fifteen were found to hold this view. Faced with the proposition that 'enough new oil will be found so that it can remain a primary source of energy for the foreseeable future', only 22% agree across all countries. Around 70% support the view that 'oil is running out and it is necessary to make a major effort to replace oil as a primary source of energy'. Most think the price of oil will go much higher, and Americans believe their government is behaving as though crude won't run out. Nearly 15,000 people were surveyed by WorldPublicOpinion.org in countries representing 58% of the world population, including the US, the UK, Russia, India and China. Only in Nigeria did a majority (53%) think that governments can rely on oil supply far into the future.[12]

Such opinions are not limited to the general public, it seems. A May 2007 survey showed a huge majority of oil executives in favour of renewables investment because of declining reserves. Polling 553 financial executives from oil and gas companies, KPMG found that 69% say at least half of government funding into energy should be directed at renewable sources. 25% of respondents think that three-quarters of funding should be spent this way. In the same poll, 82% cite declining oil reserves as a concern. 60% believe the decline is irreversible.

But the popularity of renewables is, at present, not always matched by a willingness to pay for them. In February 2008, an FT/Harris survey in the

UK, Germany, France, Italy, Spain and the US showed a clear majority of Western Europeans fear Russia as an unreliable energy supplier. However, a majority also said they were not willing to pay extra for renewable supplies. Of those who would, the majority would only pay 5%.[13]

This means that people are looking to government for leadership on bringing renewables into play so climate change can be mitigated and energy insecurity countered. It also places heavy emphasis on the economics of solar, a subject we examine in Chapter 5.

In conclusion, we argue that the case for switching entirely away from fossil fuels is strong and becoming stronger, and that solar technologies can be key to that revolutionary switch – and even the backbone of it. Before looking at what's needed to make this happen, however, let's first explore how solar technologies actually work.

Solar homes in Maidenhead, UK

Chapter 4
SOLAR TECH

How Solar Technologies Harvest Energy From the Sun

This chapter considers the processes and technologies needed to turn light into power: first for solar photovoltaics, and then for solar-thermal. It then considers the relative strengths and weaknesses of these two ways of harvesting the sun's power. The aim is to provide a good technological background for the solar case histories in the following two chapters.

The photovoltaic effect

The photovoltaic effect is the ability of a semiconducting material to convert light directly into electrical energy. (A semiconductor is a solid with an intermediate electrical conductivity somewhere between a conductor and an insulator.) The photovoltaic phenomenon was discovered by Edmond Bequerel, who noticed when experimenting with batteries in 1839 that the amount of electricity produced would vary with light. In 1876 William Grylls Adams and Richard Evans Day discovered that selenium produces electricity when exposed to light. Selenium solar cells were thereafter used for extremely low power applications, such as photographic exposure meters, right up until the 1970s. Although there was a lot of scientific research on the photovoltaic effect in the first half of the twentieth century, including a paper by Albert Einstein which won a Nobel Prize, the breakthrough happened in 1954 at Bell Laboratories. Scientists working there made the first silicon photovoltaic cell, increasing the energy output to a much more useful level.

This groundbreaking cell achieved 6% efficiency: that is, it converted 6% of the potential energy in the sunlight into electricity. Over time, the technology was developed further to achieve 15% and then over 20% efficiency. Crystalline silicon solar cells in use today commonly have an efficiency of around 16%, though the best performers reach 25% in the laboratory. (Note that the cells are used in panels, also known as modules, as described later in the chapter. These have slightly lower efficiencies than the cells they contain, because not all the panel area is covered in cells, and because there is slight variability in performance between individual cells.)

Initially, the silicon photovoltaic cell struggled to find a market, but in the late 1950s and 1960s it became the default power source for spacecraft and satellites. The next big step forward came from Elliot Berman's Solar Power Corporation, where the production cost was reduced by a factor of five using a less pure grade of silicon. This 80% cost reduction made photovoltaics a cost-competitive technology for remote areas without access to the grid. As a result, they began to be used for applications such as telecommunications. Today there is a fully established global photovoltaics industry, worth around $50 billion in 2008, which has been growing at a rate well in excess of 50% per year for the last three years – driving costs ever lower, as we'll soon see.

How a photovoltaic cell works

The basic structure of a photovoltaic cell is shown in Figure 13. Photovoltaic power generation happens when light is absorbed in a semiconductor material to release positive and negative charge carriers which are extracted from the material as an electric current. The material in most cells used today is silicon, one of the most abundant elements on earth. Silicon combines with oxygen readily to form minerals called silicates, such as quartz (silicon dioxide, or SiO_2), which is the main mineral in sand. Pure silicon can be separated from sand, in a manner described below, and this is the material used in silicon solar cells.

Generation of an electric current comes about as follows. Each silicon atom has four bonding electrons in its outer shell and, in the high-purity crystals needed to make a solar cell, neighbouring silicon atoms join together by forming electron bond pairs. Light is capable of breaking these bonds, freeing an electron, and raising the electron to an energy level where it can move through the silicon crystal. The free electron leaves behind a positively charged hole in the crystal lattice which will also move around. The electrons are negatively charged and the holes are positively charged. Their motion is random but if the electrons can be

How a crystalline PV cell works

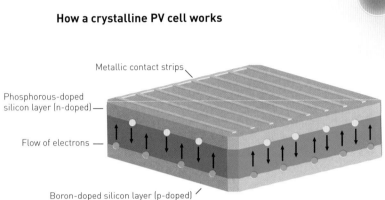

Metallic contact strips

Phosphorous-doped
silicon layer (n-doped) —

Flow of electrons —

Boron-doped silicon layer (p-doped)

Figure 13: How a photovoltaic
cell works.

forced to move in one direction and the holes in the opposite direction then they will form an electric current. The breakthrough at Bell Labs was to find a way to modify the silicon to ensure this separation occurs.

If impurities are introduced in controlled quantities into the crystal lattice – a process known as 'doping'– the crystal structure can be primed for electricity generation. A phosphorus atom has one electron more than silicon in its outer shell, and a boron atom has one electron less. This makes them perfect for the job. The cell can be thought of as a sandwich of two layers, one phosphorus-doped and one boron-doped. At the junction between the phosphorus-doped silicon layer (the negative-doped or n-doped layer, as the technologists call it) and the boron-doped layer (the positive-doped or p-doped layer) there is an electric field which makes the electrons produced by the sunlight go one way and the holes the other. This makes a current, which can be picked up in a circuit if wires are connected to the cell. Because the electrons are at higher energy levels than the holes, a voltage is created which drives the electrons and holes round the circuit in opposite directions and gets them to do work. For power, both current and voltage are required.

Light needs to pass through into the cell, and so in most crystalline silicon cells the metallic contacts on the front are thin strips, usually screen-printed as a grid onto the cell. At the back contact, in most crystalline silicon cells, the metallic layer can cover the whole surface.[1]

What we have just described is the process as it happens in a crystalline silicon solar cell. In thin-film photovoltaic cells the basic physics is the same, but charges are separated in a different way, which we describe in the section on thin-film cells below. First let us consider the manufacturing process for crystalline silicon cells, which make up around 90% of the global photovoltaics market today.

Crystalline silicon: from sand to solar

The photovoltaics industry can best be thought of as a chain of industrial activities, each quite different but entirely dependent on the one before. Let's take a look at these stages in turn, starting at the top of the chain.

Solar feedstock

The journey begins with the melting of sand to provide feedstock for the industry. A mix of silicates and carbon is heated to over 1,900°Celsius in a furnace. The carbon bonds with the oxygen from the silicates and the result is liquid silicon plus carbon dioxide or carbon monoxide. The liquid silicon left at the bottom of the furnace is 98% pure and known as metallurgical-grade silicon. Usually, if this is made into solar cells, it must first be purified into solar-grade silicon, which is 99.99999% pure.

To purify silicon it is converted into an intermediate silicon-rich gas. The most common is trichlorosilane, although silicon tetrachloride and silane are also used. The gases are blown over 'seeds' of pure silicon at high temperatures. The seeds can be rods or smaller particles which drop through the heated silicon-rich gas. Either way, the gases deposit so-called solar-grade silicon around the seeds.

The Siemens process is one of the most common methods of making solar-grade silicon. In this process the trichlorosilane gas is heated to 1,150°Celsius and exposed to high purity silicon rods which grow as the silicon is deposited onto them. Another method is to use a so-called fluidised bed reactor, which makes use of small particle-type seeds.

Silicon feedstock production is an energy- and capital-intensive business. It costs between $500,000 and $1 million to produce enough silicon for each megawatt per year of solar capacity, and it takes one-and-a-half to three years to build each new plant. This means that historically this part of the value chain has been the domain of big chemical companies. Until recently, there were only half a dozen main players. But with the rapid growth of the solar photovoltaics industry, and the huge influx of investment this has entailed, the situation is changing. Silicon

production for the solar industry exceeded silicon production for the semiconductor industry for the first time in 2006[2]. By 2008, global solar feedstock production amounted to roughly 70,000 tonnes. (As a rule-of thumb, it takes 7.5 grams of silicon per 1 watt of solar-cell power, so 1 tonne of silicon can produce cells that generate around 130 kilowatts of solar photovoltaic power.) The amount of solar feedstock production expected in 2009 by the more optimistic industry analysts is in excess of 120,000 tonnes.[3] Sixteen companies account for 85% of that anticipated production.[4]

Wafers

The solar-grade silicon needs to be made into wafers. There are two types of wafer: polycrystalline (sometimes also referred to as multicrystalline) and mono-crystalline. The solar-grade silicon pieces, which can include off-cuts and breakages from other parts of the production process, are melted in a large crucible, with boron generally added to start the doping process. For polycrystalline silicon, the molten silicon is cooled into a large block which is then sawn into bricks with a square cross section, typically 150 millimetres by 150 millimetres. Wire saws are then used to slice the polycrystalline silicon brick into very thin wafers, typically 150–250 microns thick. A micron is a thousandth of a millimetre, so the wafers are around a quarter of a millimetre each – or twice the width of a human hair.

The monocrystalline approach, by contrast, involves drawing a single crystal of silicon from a bath of liquid silicon, usually using the so-called Czochralski process. This results in an ingot (a block of silicon) with a circular cross section. Before the ingot is sawn into wafers the edges are trimmed to produce a pseudo square cross section that results in

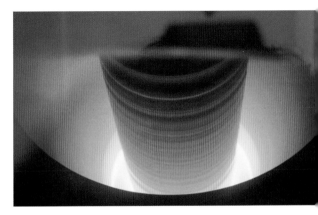

The production of a monocrystalline solar cell PV cell - the molten silicon being drawn from the crucible (above) and the cooled ingot before slicing into wafers (left).
(Images: Trina Solar)

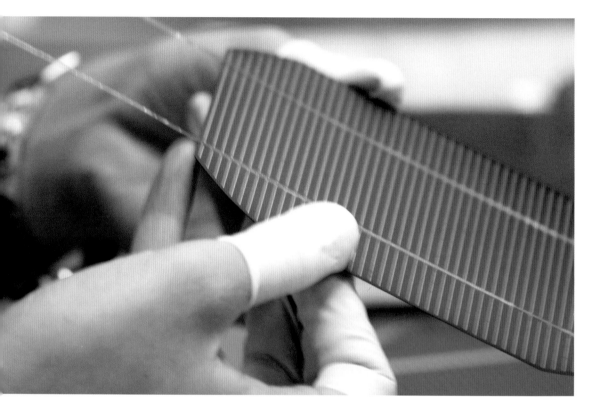

A completed solar cell, before assembly into a module.
(Image: Trina Solar)

wafers which appear to have had the corners knocked off.

Whilst solar cells made from polycrystalline silicon are generally less efficient than those made from monocrystalline silicon, the fact that the cells are truly square and can be packed together in a module without any wasted space means that the modules generally have the same power density. One company, Evergreen Solar, has developed a technology that has no need for sawing of ingots. Two thin ribbons of monocrystalline silicon are pulled slowly from a crucible, and these can be cut directly into wafers.

Wafer production requires capital expenditure of between $300,000 and $800,000 per megawatt, and it takes nine to eighteen months to build a new plant, with a further six months or so for equipment installation and set-up. The range in these figures (and those quoted for cells and modules below) is due to differences in technology, geography, vendor and level of automation.[5]

Wafers to cells

The silicon wafers are now ready to be converted into solar cells. The first stage is to prepare the surface of the wafer following the sawing. The cell is etched to clean up the surface and increase light capture. The silicon is then doped on the light-facing side with phosphorus, and an anti-reflective coating is added. Without this, the silicon would reflect up to 35% of sunlight, making it much less efficient. Finally, electrical connectors are printed on to the top and bottom of the cell, and conductive ribbons are attached to allow the cells to be connected together.

The typical efficiency of a monocrystalline cell today is 16%. The most efficient cells available commercially at the time of writing are produced by a manufacturer called Sunpower, which has designed a cell without metallic contacts on the front. These aesthetically pleasing black cells reach an efficiency of 22%. In Sunpower laboratory work, efficiency has been lifted to 23.4%. The world laboratory record for cell efficiency at the time of writing is 24.7%, held by the University of New South Wales.

Cell production requires capital expenditure of between $200,000 and $600,000 per megawatt, and it takes twelve to twenty-four months to build a new plant, with a further three to nine months for equipment installation and set-up.[6]

Cells to modules

A silicon solar cell will have a voltage between the front and the back of the cell of just half

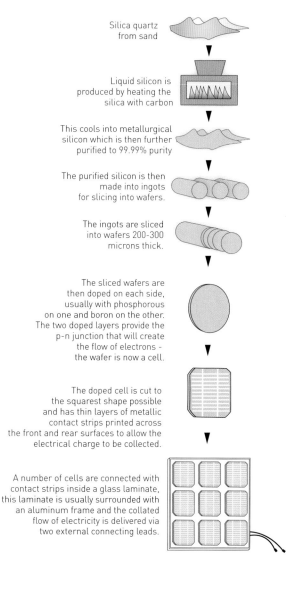

Silica quartz from sand

Liquid silicon is produced by heating the silica with carbon

This cools into metallurgical silicon which is then further purified to 99.99% purity

The purified silicon is then made into ingots for slicing into wafers.

The ingots are sliced into wafers 200-300 microns thick.

The sliced wafers are then doped on each side, usually with phosphorous on one and boron on the other. The two doped layers provide the p-n junction that will create the flow of electrons – the wafer is now a cell.

The doped cell is cut to the squarest shape possible and has thin layers of metallic contact strips printed across the front and rear surfaces to allow the electrical charge to be collected.

A number of cells are connected with contact strips inside a glass laminate, this laminate is usually surrounded with an aluminum frame and the collated flow of electricity is delivered via two external connecting leads.

Figure 14: The production process for a crystalline silicon solar module.

a volt: less than half that of an AA battery. In order to provide a useful voltage from a panel, many cells need to be connected together in series – like a chain, one after the other. To protect the connected cells from contamination they must be encapsulated in a waterproof package. The most common form of encapsulation is a sheet of tempered glass on the front (to allow light through) and a protective film on the rear. This sandwich is held together with adhesive which is cured when the whole package thus far (called a laminate at this stage of the process) is heated and pressed in a solar laminator.

The final step is to put a frame around the laminate to protect the edge of the glass and provide a structure which allows the panel to be bolted to a suitable support structure. This is usually made of aluminium or stainless steel. Alternatively, the laminate can be inserted directly into a specially designed frame to make a roof tile, or some other element that can be attached to buildings.

Module production requires capital expenditure much lower than the earlier stages: between $40,000 and $230,000 per megawatt. It takes nine to eighteen months to build a plant, with a further three to nine months for equipment installation and set-up.[7]

Quality control is vital before modules are released to market. Each module undergoes testing in the factory for its electrical properties. There are many testing sites around the world where inferior manufacturing can quickly be exposed, and the commercial downsides for a manufacturer who produces a module that doesn't 'do what it says on the tin' are severe.

Thin-film photovoltaics

Thin-film solar cells have caused a lot of excitement, and received a lot of investment, over the previous few years. These cells exploit the photovoltaic effect in the same basic way that crystalline silicon cells do, but the manufacturer uses less material and requires lower processing temperatures to reach a finished product. Thin-film solar cells are made by building up super-thin layers of materials on a suitable substrate. The final layer can be as thin as a micron: 200–300 times thinner than a crystalline cell. The energy and material costs are lower and the potential for rapid scale-up of mass production is therefore

Thin film PV module manufacturing at First Solar's plant in Frankfurt, Germany.
(Image: First Solar)

greater. At present, the efficiency of thin-film PV modules (5–12%) is lower than crystalline PV, but thin-film cells made in laboratory conditions have achieved efficiencies which rival crystalline silicon cells.[8]

There are three main thin-film technologies currently vying for commercial success: thin-film silicon, cadmium telluride, and copper indium gallium selenide (CIGS).

Thin-film silicon

Silicon can be deposited onto glass or metal substrates from gas in a vacuum chamber. Doing so reduces the amount of silicon required in a solar cell by around 99%. Silane, one of the gases used in the manufacture of solar-grade silicon, is the raw material for these thin-films.

Depending on the rate at which the silane deposits the silicon on to the substrate, the silicon thin film can have different structures. The most common form is amorphous silicon, but so-called protocrystalline and nanocrystalline structures can also be created. These different structures can exploit the energy in different wavelengths of light, so by building up multiple layers it is possible to use a greater part of the solar spectrum.

Some manufacturers, notably Unisolar, have managed to deposit thin-film silicon-based devices with three-stacked cells onto stainless-steel foils. The manufacturing cost of this 'triple junction' technology is approaching $2 per watt.[9] The result is a flexible, lightweight product which opens up new architectural possibilities as well as portable applications.

Cadmium telluride

In these cells a compound semiconductor made from cadmium and tellurium is used as the light-absorbing material. Cadmium telluride thin-films have reached an efficiency of 11% in modules on sale, and

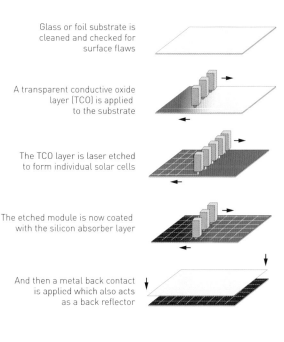

Glass or foil substrate is cleaned and checked for surface flaws

A transparent conductive oxide layer (TCO) is applied to the substrate

The TCO layer is laser etched to form individual solar cells

The etched module is now coated with the silicon absorber layer

And then a metal back contact is applied which also acts as a back reflector

The module is flipped over and mounted glass side up in an aluminium frame, connecting leads are applied which allow the electricity flow to be used

Figure 15: The production process for a thin-film solar cell.

16.5% in cells in the laboratory, with experts projecting efficiencies of 20% long term – higher than regular silicon modules today. The US National Renewable Energy Laboratory believes cadmium telluride offers the scope for the lowest manufacturing costs amongst the various thin-film technologies. First Solar, by far the biggest thin-film manufacturer to date, has recently lowered manufacturing costs below $1 per watt for the first time ($0.98), and is on track for still lower costs as it scales up.[10]

Concerns about the heavy metal content of these cells have affected the market uptake historically. However, First Solar has put in place a recycling process that recycles the cadmium and tellurium from used modules back into their manufacturing process (glass goes to a glass manufacturer for recycling). This takes away many of the environmental concerns.

CIGS

Thin-film cells can also be made with copper, indium and selenium, resulting in so-called CIS cells. If the mix also contains gallium, you get CIGS cells.

CIGS is one of the most exciting thin-film technologies, reaching an efficiency of 20% in the laboratory today, with 25% expected in the long term. So far, however, efficiencies in modules on sale are much lower – around 11%. This large efficiency gap between factory and lab – larger than the gap for crystalline and the other main thin-film variants – is because of the difficulties of getting four elements to deposit in a uniform film.

In Silicon Valley, as technologists and investors alike shift their interest from the digital revolution to the solar revolution, companies have attracted high levels of investment to CIGS technology. Among them, Nanosolar has said it will be able to cut manufacturing costs for solar cells to as low as 30 to 35 US cents per watt. This, plus the suitability of CIGS technology for 'roll on' manufacturing (where the thin-film layers are deposited on the substrate as it is unrolled, almost like sticky tape), are among the reasons why this technology has attracted such high levels of investment.[11]

Next-generation solar-cell technologies

There is a wide family of solar-cell technologies that collectively are known as 'next generation'. Some are exotic and still very much at the primary research stage, whilst others are in market and their manufacturers are busy raising capital for huge production plants. The core promise of next-generation solar cells is an additional step down in cost beyond what's possible with silicon thin-film, cadmium telluride or CIGS. This cost reduction comes from a combination of reel-to-reel processing and the use of fundamentally cheaper core materials than used for other photovoltaics. However, most of these next-generation technologies, and certainly those that have been commercialised, suffer from two shortcomings. The first is that they are of low efficiency, only 5–10% at best. The second is that they have a relatively short lifetime, of the order of a few years rather than decades.

One of the technologies already in market is the dye-sensitised solar cell. This is a type of thin-film technology based on a very cheap semiconductor material, titanium dioxide, which is also used in toothpaste and paint, with a minute amount of liquid electrolyte embedded

Reel to reel solar: thin film technology at Konarka's manufacturing plant.
(Image: Konarka)

into the film. Its advantages are relatively high efficiency (over 10% for the best cells), low cost and mechanical robustness. Because the cells are translucent, they can in principle be used as photovoltaic windows in buildings. In addition, a dye-sensitised solar cell performs very well under conditions of low or diffuse light. However, there are issues to be resolved around the long-term stability of the film before such cells can be expected to be a credible substitute for silicon photovoltaics in permanent external installations.

Flexible thin film PV
integrated into a bag
by Konarka.
(Image: Konarka)

Another technology that is currently being commercialised is the organic photovoltaic cell. In this, the active layer is made of a mixture of compounds based on carbon. Organic photovoltaics are sometimes referred to as 'plastic PV' because they can be deposited from solution by high-volume coating techniques like conventional plastic films. They have many advantages, being very cheap, flexible and light. Although the efficiencies of organic solar cells are still low at 5–6%, they have improved rapidly during this decade and efficiencies over 10% are expected. However the relatively low efficiency and durability mean that the first applications are likely to be in light-duty systems such as integrated chargers for backpacks or tents.

Cost-reduction: the holy grail for photovoltaics

Solar manufacturers regularly announce small increases in the efficiencies of their cells, and many people think as a consequence that this is the main way costs will be reduced in the photovoltaics industry. But in fact at every step of the photovoltaics value chain, industry players are striving to reduce costs in other ways, too. These various approaches will be key to achieving 'grid parity' – the point where solar is cost-competitive with traditional electric power – in all markets. In many ways, this is what solar technology development is all about. Let us consider first the developments in the crystalline silicon segment of the industry, then trends within the industry as a whole as they impact on cost and price.

Squeezing cost from the chain

In the silicon feedstock industry, one route to cutting costs is to find ways of using so-called upgraded metallurgical-grade silicon (UMG) – silicon that has not been purified to solar-grade – in the making of viable cells. Early efforts by manufacturers show that this is feasible with only a slight reduction in the efficiency of the final cell, compared to a potentially large

reduction in manufacturing costs. This is an area of innovation to watch, as the solar industry grows and evolves.

The fifty or so companies making wafers for solar cells pay more than 95% of their cost of goods for the silicon ingot, so their focus is on making best possible use of it. They are aiming for larger wafers, thinner wafers, thinner saw wires and faster throughput. The risks they and the feedstock manufacturers face start with the fact that melting silicon can only be done at scale, because it is capital-intensive, vulnerable to interest-rate changes and holds the potential for stranding of assets. An added risk for wafer manufacturers, if they are not also manufacturing their own silicon feedstock, is supply. On the benefits side of the equation, though, margins have been high for many years because demand has been growing so fast at the user end of the value chain. That is why so many new companies have been buying their way into the feedstock business in recent years.

The sixty-plus companies making cells pay some 60–70% of their cost of goods for the wafers, 10–20% for the various pastes needed to make a cell, and up to 5% for chemicals. They are aiming mainly for higher throughput, and larger wafers. They face the risk of overcapacity as a result of supply difficulties, a problem many companies have experienced in recent years because demand has been so high. High demand also leaves them subject to high pricing by feedstock and wafer manufacturers. For this reason, some cell manufacturers have elected to set up wafer and feedstock operations of their own.

Hundreds of companies are involved in module manufacturing. Their cost of goods is around 60–70% in the cells, 10–15% in the glass, 10–15% in the frame, and 10–15% in the adhesive needed for lamination. To trim costs they aim for larger cells, less material usage, and higher automation. They too face the threat of overcapacity, and the crowded market means pricing pressure.

Finally, the modules need to be installed, and thousands of companies are involved at this end of the value chain. Their cost of goods is dominated by the module price, which to the great joy of installers is dropping fast at the time of writing. To trim costs they aim for standardisation of systems, faster installs, and increased system integration. The main risks in this business involve the low barriers to entry in the market.

The average price per watt of a solar-photovoltaic module at the factory gate in 2008 was $3.98, and the average wholesale price of that module was $4.18. The average price of a fully installed system was $7.56. Within just two years, analysts expect the average factory-gate price to have reduced to $3.17, and the average fully installed price to have dropped to $6.05.[12] These, it should be stressed, are average figures. They serve as benchmarks for comparison, because lower figures can and are being achieved. (Note in the discussion which follows the difference between manufacturing cost and sales price: the manufacturing cost is what the manufacturer pays to produce the product, the sales price for the module includes the profit margin, and the fully installed price also includes the margin charged by the installer.)

Capturing the whole chain

As the photovoltaics industry grows, companies beginning life in one sector of the value chain will expand not just in volume and geographic spread but also in what is called vertical integration. That is, they will begin to operate in different sectors of the value chain. The advantages are obvious. In particular, they cut their supply risks, and they reduce their costs because they do not have to pay a margin to the supplier. Equally there are risks, as there are with any business expansion.[13]

The process of vertical integration is already happening. Companies that started life making cells are moving both upstream and downstream. Companies that began life as installers are moving upstream. Silicon companies are moving downstream. Some companies aspire to spread their own operations across the entire value chain, notably the German company Solarworld.

Solar PV manufacturers are scaling up fast. To take one example among crystalline manufacturers, Chinese company Suntech began manufacturing only in 2002. By the end of 2007 it was the third biggest manufacturer of cells in the world, with a capacity of 350 megawatts per year and production of 336 megawatts. By the end of 2008 its capacity was heading for a gigawatt (1,000 megawatts). A year's output from such a plant could generate power on the scale of a typical nuclear power plant.

Most importantly, with size comes further cost reduction, and it is this fundamental relationship – rising scale, falling manufacturing cost – that is driving much of the investor interest in solar. While gas, oil and coal costs have generally inflated in recent years, the cost of producing solar

energy has fallen roughly 20% every time global manufacturing capacity has doubled (about once every two years at the moment). Suntech has a price target for its integrated PV systems of $4 a watt by 2012, at which level electricity could be generated for 18–20 US cents per kilowatt-hour, compared to 19–36 cents at the time of writing.

Many experts believe that thin-film offers even more opportunity for cost reduction. First Solar is currently the largest manufacturer of cadmium telluride solar cells, with over 400 megawatts of manufacturing capacity and production costs reportedly approaching $1 per watt. The company plans to be selling modules at around $1.25 per watt by 2012, which would put the cost of a kilowatt-hour at 8–10 US cents, easily cheaper than conventional electricity by then.[14]

Applied Materials, a company that grew to giant size during the digital revolution, recently decided to enter the photovoltaics business in a big way. It doesn't manufacture actual solar panels, it manufactures factories – 'fabs' in the industry jargon – for the fabrication of panels. In 2008, the company received its first order for a gigawatt-a-year fab. Applied Materials is betting that the solar market will bifurcate into two main sub-markets. A space-constrained rooftop market will mostly use increasingly high-efficiency crystalline modules. Meanwhile, solar PV farms (see p.73) will mostly use large-area thin-film silicon modules mounted on simplified mounting structures. It is in the business of providing clients with both types of factory, and in doing so is adopting equipment from other much larger industries of which it has experience, such as the coated-glass and flat-panel-display businesses.

Applied Materials and its main competitor, Oerlikon, offer thin-film factories capable of manufacturing cells at a cost of $1.27 per watt, at the time of writing. All-in-one crystalline factories are also being developed, and they are not far behind in terms of cost. Centrotherm offers a 347 megawatts-per-year crystalline-silicon factory, taking three years to build, that it says will be able to produce 16.5%-efficient cells with a manufacturing cost as low as $1.36 per watt if sited in China, and $1.57 in the US. The factory, costing $915 million, is really five sub-factories, which could be located on separate sites. The polysilicon production, with a capacity of 2,500 tonnes a year, would best be located in an area of less expensive electricity, for example, next to a hydroelectric station. The module-production facility, meanwhile, could be located closer to the final market, to cut transport costs.[15]

Some experts believe thin-films will become the dominant solar technology. Analysts say that thin-film manufacturing will total at least 4 gigawatts per year by 2010, which would be approaching 20% of the global photovoltaics market, up from around 10% in 2009. One European analysis counts fully 10 gigawatts of anticipated thin-film capacity by 2010, across 113 companies: 82 pursuing the thin-film silicon route, 19 focusing on CIS or CIGS, 7 on cadmium telluride and 5 on dye-sensitised cells.[16] This is certainly encouraging. But the writers of this book believe both crystalline and thin-film will have their places in the future, on the 'horses for courses' principle. Whatever the result of the race to be cheaper, the different modules have very different appearances, and for use on buildings they will appeal to different architects in different ways. It should also be noted that many companies are hedging their bets, and manufacturing both types of technology.

The eventual proportion of crystalline silicon and thin-film in the energy mix of nations will also be much dependent on wider economics, of course. This vital subject we consider further in Chapter 5. Suffice it to say here that investment-bank analysts have recently concluded that, in areas of high insolation, thin-film systems installed at a total cost of $4 per watt are already at grid parity, even for today's gas prices.[17] With the kinds of manufacturing costs cited above for all-in-one factories, and the prices cited for cadmium telluride and crystalline silicon at scale, plenty of room is left for profit. Even without all-in-one factories, both crystalline and thin-film manufacturers expect to be at grid parity in most or all markets within just a few years at the most.

Solar-Fabrik manufacturing facility, Germany.
(Image:© Solar-Fabrik AG, Freiburg.)

Solar farm under construction,
Masdar, Abu Dhabi.
(Image: Environmena)

Solar-photovoltaic farms

The quickest route to grid parity is to array solar panels in rows on scrubland in sunny locations. Such installations are known as solar-photovoltaic 'farms'. Direct current flows to a DC junction box and then to an inverter where the DC is converted to AC power. In order to reduce power losses the AC voltage will be stepped up to a high level through a transformer and connected into the local power-distribution grid, as shown in Figure 16.

The number of solar-photovoltaic farms commissioned has been accelerating rapidly since about 2004 and looks set to continue. More than 90% of the PV manufactured worldwide in 2008 will be used for solar PV farms. Today the farms tend to vary in size from around 5 up

Rows of thin-film (dark) and crystalline (light) modules in the Masdar Solar Farm, Abu Dhabi.
(Image: Environmena)

to around 20 megawatts. At the time of writing, Spain has over 300 farms of 1 megawatt and larger, providing a total power of 1.45 gigawatts.[18] At less than 2% of national electricity supply (86 gigawatts), this is barely a beginning. There is a 60 megawatt PV power station under construction in Spain, and Californian utility Pacific Gas and Electric plans an 800 megawatt solar farm: a single installation almost twice the current generating capacity of all photovoltaics currently installed in the US (473 megawatts). As big as a natural-gas power plant, the farm will be located in San Luis Obispo County. Its 250 megawatts of crystalline silicon and

550 megawatts of amorphous-silicon thin-film modules will provide 1.65 billion kilowatt-hours of electricity a year, enough for nearly a quarter of a million homes, peaking in the afternoon on the sunniest days when electricity demand is at its highest.[19] The first arrays from the plant will begin generating power in 2010, and the whole farm will be complete by 2013, according to Pacific Gas and Electric's plan. Analysts expect the first solar electricity from the plant to be priced as low as 12 cents per kilowatt-hour. Californian combined-cycle gas power plants currently generate electricity at around 10 cents per kilowatt-hour. But gas prices seem set to rise going forward, while solar prices will fall further. Moreover, a Renewable Energy Portfolio introduced by the Californian state government requires utilities to generate 20% of their electricity from renewables by 2010.[20]

Some solar-photovoltaic farms employ tracker technology, where the modules are mounted on stands that move on one or two axes to maximise the amount of solar radiation captured. The largest of the tracker farms at the time of writing is 19.6 megawatts, in South Korea. The introduction of moving parts increases the price of the system, and introduces more scope for maintenance downtime, but of course the increased intensity of light on the module increases yield, and therefore the amount of electricity generated. There are currently nearly 100 different types of tracker on the market. It is still too early to say whether or not such systems will develop a decisive price advantage over static systems.[21]

A solar photovoltaic farm

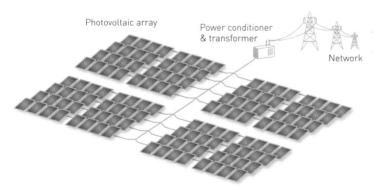

Figure 16: The basic elements of a solar-photovoltaic farm.

Solar-photovoltaic concentrators

Another route under active investigation for lower-cost photovoltaics involves concentrating the sun's rays, using mirrors and/or lenses, and focusing them on small strips or patches of photovoltaic material. Since mirrors and lenses are much cheaper than solar cells, cost can in principle be cut deeply this way. Sunlight can be concentrated more than 1,000 times with existing technologies. The photovoltaic cells in concentrators can be made of traditional silicon, or – because such small areas are needed – the more expensive high-efficiency cells made of gallium, indium and germanium that are normally associated with space applications.

Concentrator companies have achieved efficiencies of over 40% in the laboratory, almost double even the best crystalline silicon cells operating without concentration.[22] Concentrated PV module efficiencies of 28.5% have already been achieved in outdoor testing and this figure is expected to rise with improvements in manufacturing quality.

Two technical challenges are: ensuring that the concentrators remain at 90° to the sun at all times, and the fact that the cells become very hot. These issues may prohibit significant use of concentrators on buildings and pose challenges for larger-scale applications.

However, over a billion dollars of investment has gone into research and development in the last few years and nearly 200 megawatts of concentrated PV panels will be installed in 2009. Manufacturing plants are being constructed, alliances with tracker

A solar concentrator array.
(Image: Concentrix)

companies being forged,[23] and executives from companies deploying concentrators expect to be generating cheaper electricity than conventional photovoltaics by 2010.[24] We will see.

A solar-photovoltaic concentrator module

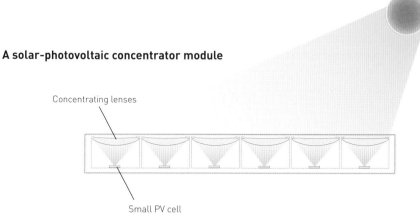

Concentrating lenses

Small PV cell

Figure 17: The basic elements of a solar-photovoltaic concentrator.

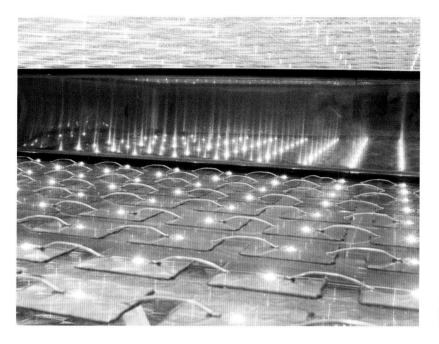

The interior of a Concentrix solar concentrator module. (Image: Concentrix)

Solar-thermal at low temperatures

The term solar-thermal describes any method used to harness the energy of the sun in the form of heat. Some technologies have been in use for over 2,000 years, whilst others are on the cutting edge of engineering and materials science. Broadly speaking, all solar-thermal technologies can be placed into two groups according to temperature and intended use. Low-temperature solar-thermal technologies are used to heat air or water for direct use. High-temperature solar-thermal technologies are used to heat oil or make steam in order to drive some form of turbine so as to generate electricity. Let's look first at the lower-temperature solar-thermal approaches.

Passive solar

Air can be warmed by the sun and then used directly to warm the interior spaces of a building. By clever use of airflow, the same warmed air can also be used to act as a solar chimney to extract hot air from a building and cool it down. The ancient Greeks began the first codification of this so-called 'passive' approach to solar. As Socrates put it, quite simply: 'In houses that look towards the south, the sun penetrates the portico in winter.'

It is also possible to warm water simply by passing it through a pipe exposed to the sun. This method is cheap, but requires a large surface area to be effective and only creates a relatively small temperature rise. One common use is for warming swimming pools, where the actual pool water is passed through a labyrinth of pipes. To produce water hot enough for taps and showers, a glazed solar-thermal collector working at a higher temperature gives much better results.

Solar-thermal collectors

Glazed solar collectors have been around since 1767, when the Swiss scientist Horace de Saussure invented one designed to reach cooking temperatures. Solar-thermal technologies didn't really catch on, however, until the end of the next century, when, in 1891, the first patent was granted for a solar water-heating device. They rapidly gained popularity,

being much better than wood- or coal-burning stoves for water heating. Many towns did not at this time have electricity, and where coal gas was available it was very expensive. In 1897, a third of the homes in Pasadena, California had water heated by the sun, and by 1920 tens of thousands of systems had been sold, particularly in California and Florida. However, the history of solar energy has long been tied to the fortunes of oil and gas. The discovery of natural-gas resources and the reduction of electricity prices caused a huge decline in the solar-thermal industry in the early years of the twentieth century.

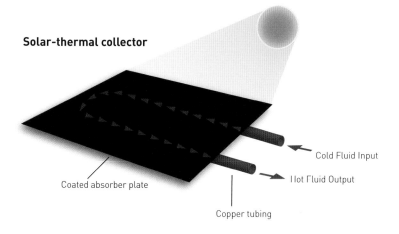

Solar-thermal collector

Coated absorber plate

Copper tubing

Cold Fluid Input

Hot Fluid Output

Figure 18: The basic elements of a solar-thermal collector.

A typical system is based on a 'flat-plate' glazed collector panel on the roof of a building. The sun's energy passes through the glass to heat up a specially treated metal plate. The key to the technology is the nature of glass, which is transparent to visible light but opaque to the long-wave radiation re-radiated from a solar collector (or interior of a building) behind it. This means that the system behaves like a small greenhouse, with heat from the sun going in faster than it radiates out, resulting in increasing temperatures. The temperature of the water can be varied in different designs of collector by either increasing the proportion of radiation absorbed (for example, by double-glazing the front of the collector), or decreasing heat lost from the system (for example, by removing insulation on the non-solar side).

Behind the metal plate are pipes filled with water. These transfer the heat to the building's hot-water system. Only a little water can be heated at a time, meaning a small pump is needed to move water into the storage tank. To get around the problem of the water either freezing in summer or boiling in winter, some systems heat water indirectly by using another fluid, such as an oil or antifreeze solution. This has the added benefit of reducing corrosion. Another option is a 'drainback' system, where the heat-transfer fluid is emptied into a container when the system is not in use.

A solar-thermal flat plate collector array in the UK.

The performance of a glazed system can be improved by removing the air between the glass and the treated collector plate, thereby eliminating heat loss by convection. For manufacturing reasons, this type of system is produced in the form of tubes, hence the name 'evacuated tube collector'. Flat-plate collectors can heat fluids to a temperature of around 80° Celsius. Evacuated-tube collectors can heat them to 100° Celsius and above. Evacuated tubes perform better in low light and low temperature than flat-plate collectors, and they take up less space. On the other hand, they are also more costly and more fragile.

In most buildings there is a hot-water boiler fuelled by gas or oil, or a hot tank powered by electricity. The solar-thermal hot water system is used to provide a carbon-free substitute. In a typical home in the UK, such a system can provide about 50–75% of the total annual hot-water requirements (all in summer, and enough in winter usefully to top up the hot-water tank). It is also possible to use this type of system to supplement a combined central heating and hot-water system. In this set-up, the solar collector can provide about 15% of the total heating requirement for a well-insulated home in a country such as the UK.

Storing the heat collected with a solar-thermal system is relatively simple and inexpensive compared to storing electricity created from photovoltaic panels. Hot water keeps in an insulated hot tank overnight, and it's even

possible to create a larger seasonal store, capable of harnessing summer heat to give winter warmth. However, in all but a few cases, solar-thermal energy in buildings will need to have a supplementary back-up system for either water heating, space heating, or both.

The great majority of solar water heaters in use today are in residential buildings and on the kilowatt scale. However, solar water heating is also used on the megawatt scale, for district heating. One of the largest such systems in the world, at Marstal in Denmark, is 12.8 megawatts of collectors, covering 18,300m^2 of land. This system, the largest of its kind in Europe, can meet baseload for the community in summer, and can be augmented by a pellet boiler. An oil-fired boiler is used in winter cold. By saving wood-pellet costs, the system has a seven-year payback and was installed with no subsidies.[25]

Large solar-thermal systems are increasingly used on commercial and public buildings with significant hot-water needs, such as hospitals and hotels. As we will see later on, solar-thermal systems can also be used for cooling buildings.

We'll come back to the pros, cons and costs of solar collectors and other solar technologies at the end of this chapter. First, though, let's explore how solar-thermal can be used for electricity generation.

An evacuated tube solar-thermal collector.

Concentrated solar power

Solar-thermal concentrators use some form of reflector or lens to concentrate the sun's energy onto a small area, super-heating steam or a

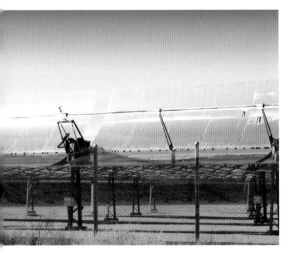

Solar-thermal concentrators in the USA

fluid that in turn drives a heat engine to generate electricity. In contrast to photovoltaics, the light energy is converted into heat rather than electrical potential energy. Also in contrast to photovoltaics, the process only works under strongly concentrated sunlight, because a high temperature is needed for the heat engine to work efficiently. This method of electricity generation first hit the headlines during the energy crisis of 1973. A project known as the Solar Energy Generating Systems, or SEGS, was started in California. There are now nine plants generating a total of 354 megawatts on the site, making SEGS the largest solar-power system of any kind in the world. In total, there are 1 million mirrors and the plant covers a total of 1,600 acres.

Since 2004, there has been a resurgence of interest in concentrated solar power and many plants are under construction or have been announced, ranging in size from around 50 megawatts to 500 megawatts and beyond. To give a sense of perspective, a typical nuclear power plant would generate on the order of 1,000 megawatts.

In order to achieve reasonable efficiency, the plants need to heat up steam to temperatures of around 600° Celsius. This requires not only a system of mirrors or lenses, but also a means of tracking the sun accurately as it moves across the sky.

There are three main types of solar-thermal farms. The first uses a series of metallic parabolic reflectors, laid out in long lines, to focus sunlight onto a pipe filled with a heating fluid such as ethylene glycol. Each line works in a similar way to the bar and reflector in an electric fire, but in reverse. The parabola are aligned north–south and rotate along their length to track the sun as it moves through the sky from east to west. The pipe at the centre of each trough runs to a heat exchanger, where the hot fluid turns water into steam that can drive a turbine, as shown in Figure 19.

Another method is to use a dish reflector. This is like a giant satellite dish, which needs to move in two axes to track the sun accurately. It is a large, high-precision piece of equipment, but has the advantage of being relatively compact, and is also capable of generating very high temperatures, which leads to greater efficiency.

The third technique is called a 'power tower' and is in effect a hybrid of the two systems. A number of mirrors are mounted in a field each with its own two-axis tracking system. The combined mirror and tracker is known as a heliostat. Each mirror focuses the sun's energy onto a single point at the top of a central tower. Again, very high temperatures are achievable along with the associated efficiencies.

The complexity of the mechanics of dish reflectors and power towers mean that they are today often more expensive than parabolic reflectors. However, there is much scope to innovate around the tracking mechanisms, and in the medium term it is widely expected that they will become the cheaper solution.

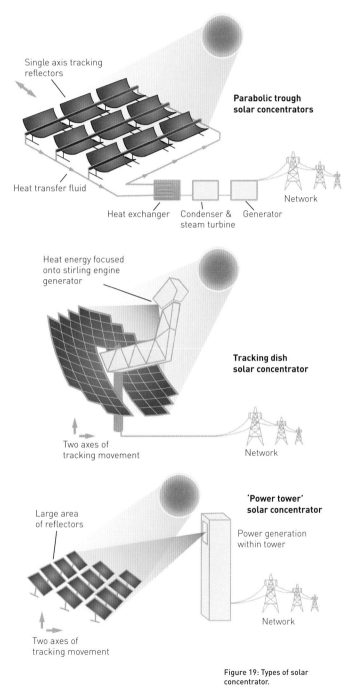

Figure 19: Types of solar concentrator.

Solar photovoltaics and solar-thermal electric: strengths and weaknesses

All energy technologies have strengths and weaknesses. Solar is no exception. Before considering the relative pros and cons of photovoltaics and solar-thermal systems, however, it should be stressed that both have very important roles to play. Proponents of one system or the other have a tendency to sing the praises of their favoured technology in a way that can lead to a 'PV versus thermal' debate. This either/or approach isn't helpful. We are going to need both photovoltaics and thermal technologies in abundance in order to bring total solar capacity up to numbers measured in terawatts rather than the mere gigawatts that we have today. (At the time of writing, the total is roughly 186 gigawatts, made up of around 15 gigawatts of solar-photovoltaic electricity,[26] just 1 gigawatt of solar-thermal electricity[27] and some 170 gigawatts of solar-thermal heating[28].) That being said, let's take a look at how photovoltaics and thermal-electric systems compare.

The condition of light required

The mirrors in a solar-thermal farm require clear blue skies for the sun's rays to be focused effectively (without significant scattering) onto the absorber tube or collector tower. For this reason, solar-thermal plants are not well suited to regions such as deserts (where there can often be airborne dust), or the tropics (with their rainy seasons), or near smoggy cities. In such locations, and of course in the cloudy northern countries, photovoltaics have a clear advantage. Some thin-film technologies, capturing light across a broad range of wavelengths by using multiple layers of different semiconductors, work very effectively under constantly cloudy skies. The editor of this book lived in the UK's first solar roof-tile home, which had triple-junction amorphous-silicon thin-film solar shingles on both the south- and north-facing roofs. Both were steep, and the north-facing roof felt direct sunlight only in midsummer. At the end of the first year, the north side of the roof had generated fully 60% of the electricity per unit area that the south-facing roof had. (1.6 kilowatts of Unisolar shingles, in all, generated more than 1,100 kilowatt-hours in the course of the year.)

Whereas planners of the Spanish solar-thermal power plants we consider in Chapter 5 have to study the solar irradiation conditions for several years before they finalise their operational model, and give construction of a plant a green light, a photovoltaic roof can be put up without much forethought at all, and it doesn't even have to face south, much less track the sun.

The workforce

A fixed solar-photovoltaic farm needs only a gatekeeper to look after the keys, and a maintenance worker on call for the rare occasion that anything malfunctions. The modules need cleaning only infrequently. A solar-thermal power plant, in contrast, will need some tens of workers: some to work in the control room and others to maintain the station, which has moving parts. Parabolic troughs will need cleaning much more frequently than photovoltaic panels, perhaps once every few weeks.

But in any comparison the entire value chain of an industry needs to be considered. Installing solar photovoltaics and solar-thermal on buildings is relatively labour-intensive. Averaged out, for every megawatt of photovoltaic capacity, 7–11 jobs are created. This compares to a maximum of 3 for every megawatt of wind power and around 1 to every megawatt of coal- and gas-fired generation.[29] That could be construed as an advantage for coal and gas, to the businessman who sees the world only through the lens of short-term profit. But as governments seek to rebuild the wreckage of their economies after the financial crash, they are increasingly likely to favour the idea of labour-intensive industries. We consider this further in Chapter 5, in the context of the green new deal.

Building a solar farm: from foundation pits to fitting the modules.
(Image: Enviromena)

Area required

To produce a megawatt-hour of electricity each year, a typical solar-thermal power plant might need 11 square metres of surface area, and a solar-photovoltaic plant using high-efficiency modules would be little different. The score is tied here, if space is a constraint. (But note that solar-photovoltaic farms can be arrayed across slopes, whereas solar-thermal plants can't.)

Costs and prices

For solar farms, thermal technology is currently less expensive for each unit of generating capacity installed. For the Spanish Andasol 1 plant (see p.112), the investment needed in each unit of equipment able to generate a megawatt-hour per year was €1,670 ($2,400). For a typical current Spanish solar-photovoltaic plant it is around €4,000 ($5,700). But there are caveats. For solar-thermal, all the capital has to be in place before the project kicks off. For photovoltaics, a farm can be built up in modular fashion, and the financing can be split into tranches too.

As for the future, we have considered cost reductions in the photovoltaics industry earlier in the chapter. Total solar photovoltaics cost-of-manufacturing is dropping by around 20% as production volume doubles, compared to 12% for solar-thermal. An example of one of the more recent concentrating solar power plants is the Accione Nevada Solar One, completed in June 2007 with a maximum capacity of 75 megawatts. This plant produces electricity at a competitive 9–13 cents per kilowatt hour.[30] Pacific Gas and Electric's planned investments in two solar-photovoltaic power plants generating a total of 800 megawatts are based on long-term power-purchase agreements that will of necessity be at a similarly competitive rate. Comparisons are all very well, but the main point is this: at these prices, increasingly expensive gas- and coal-fired electricity will not be able to compete with either solar-thermal or photovoltaics.

Water supply

Solar-thermal power plants require huge amounts of water to operate: around 4,500 to 5,500 litres per megawatt-hour generated. This is about the

same amount as is needed to irrigate crops such as wheat on the equivalent space occupied by the solar farm: a particular problem when you consider that solar-thermal power plants operate best in sunny areas where water will tend to be in short supply and needed for agriculture. This is not a problem faced by photovoltaics, which require no water to operate.[31]

Energy payback

One argument often expressed against solar power is that making solar-photovoltaic panels requires more energy in production than can ever be produced by the panels. If that were true, this technology would create rather than reduce carbon emissions. It's a serious charge – but is it true? In one word: no. To borrow the words of the US Department of Energy: 'Based on models and real data, the idea that PV cannot pay back its energy investment is simply a myth. Indeed, researchers [have] found that PV systems fabrication and fossil-fuel energy production have similar energy payback periods (including costs for mining, transportation, refining and construction).' For modern panels, the energy payback period is around one to four years, depending on the technology used. Assuming a life expectancy of thirty years for the panels, that would mean 87–97% of the energy generated would be emissions-free.[32] Manufacturers, of course, insist that their modules will last for much longer than thirty years. Indeed the first modules ever to be deployed have been in the field for more than forty years, and are still providing clean energy today, which makes the energy created even closer to zero-carbon. Increasing the efficiency of the manufacturing processes, as manufacturers scale up plants, will reduce the energy used in production further still.

Recently Norwegian solar giant REC has said it can reduce the energy payback for crystalline modules to less than a year, lower than any other module on the market. This it can do by using the fluidized bed reactor method of ingot production, which operates at lower temperatures than the widely-used Siemens process, meanwhile using thinner wafers, higher efficiencies, reducing the amount of aluminium and glass used in the module, and manufacturing near to sites where hydro-electric power can be used.[33]

Solar space heating and cooling in buildings

So far we've covered solar water heating and electricity generation, but the sun's thermal energy can also be harnessed for heating and cooling buildings, both by passive and active approaches.

Solar space heating

Passive solar space heating is primarily, but not exclusively, a method for night-time heating. It works on the principle of capturing the sun's energy, storing it in some way in the fabric of the building, and then releasing it to provide warmth when it is dark. An additional advantage is that, because some of the sun's energy is absorbed during the day, there will be an improvement in comfort as the daytime temperature in the building is kept lower than it otherwise might be. In its simplest form, passive solar space heating need be no more than a south-facing glazed window, and a floor made of a material that is a good thermal absorber. This type of thinking goes back more than 2,000 years and is known as a 'direct gain' system.

Better performance can be achieved from an 'indirect gain' system. In this case a thermal mass, such as a wall, is positioned behind a south-facing glass façade. By use of vents, the heat flows can be controlled to provide some natural ventilation during the day, and radiant heat at night. Such control also allows the use of a variety of other heat-storage sources, such as water in a roof pond or air in a conservatory.

Active solar space heating, in contrast, uses the sun's energy to provide carbon-free space heating whilst the sun is shining during the day. A glazed solar collector, similar to those discussed above, heats water to supplement a normal hot-water radiator system, reducing the amount of oil or gas needed by the boiler to get the water to the high temperature needed. Another method uses a device called a heat exchanger. Here, the hot water is pumped through a matrix of pipes, and air from a room is blown across the pipes. Thus the air is warmed before being returned to the room.

Systems also exist where there is no water involved. Air is simply blown from a room into a glazed roof collector, where it is warmed, and then returned to the room. These types of systems can also be used to suck hot air out of a room and provide passive solar cooling.

Solar cooling

There is something fundamentally elegant about the idea of having a solar-powered cooling system for a building. The hotter it gets outside, the greater the level of solar energy, and therefore the more cooling power available. There is also an urgent need for a low-carbon solution for cooling buildings as air conditioning takes a lot of energy, more than heating. The carbon footprint of air conditioning is rapidly rising, as the world warms and as some hot countries become more affluent. However, a lot of innovation is still required in order to come up with a solar cooling system that is technically feasible as well as financially viable.

All active cooling systems use a device called a heat pump. With a fridge, heat is pumped out of the cool box and into the kitchen, which is why the back of a fridge is always warm. Room cooling works the same way: heat needs to be pumped out and dumped outside the building. Almost all current air-conditioning systems and fridges use an electric motor to pump a refrigerant cycle.

One might imagine that solar cooling need be no more than a PV array to drive a conventional electric air-conditioning unit. However, this simply doesn't work for two reasons. Firstly you need so much PV to power any meaningful amount of cooling for a building that there is probably not enough space on the roof. Secondly, because you need so much PV, the system would be too expensive, even if the roof were big enough.

The solution is to use a special form of heat pump that uses hot water as its energy source instead of electricity and to get the hot water from a conventional solar-thermal system.

The core technology in the vast majority of such systems is called absorption cooling. This is not new, and was once common in industrial facilities that had large boilers with excess capacity during the cooling season. However, absorption cooling driven by solar hot water is still in its infancy and, whilst there are a number of successful installations, most are one-offs, and there is not yet a default system design. That said, there are also many technical developments under way to improve efficiency and reduce cost, and some companies are emerging with commercially interesting systems.

There are three main challenges to be addressed. The first is to come up with a system that will work with the type of energy that is available from a solar-thermal system: that is, hot water in the region of 60–95°

Celsius. The second is to achieve sufficient efficiency to yield meaningful cooling for a building of a given floor-and-roof area. The third is to develop a simple system that is cheap to manufacture and is reliable in use, particularly for smaller buildings. Although there are a number of technical differences at the detail level, all systems essentially do the same thing. They take in hot water from a solar-thermal system, and use the energy stored in the hot water to chill a separate body of water down to about 5° Celsius. The chilled water can then be used directly for underfloor cooling, or indirectly via air-handling systems. Because an absorption cooler normally works in cycles rather than continuously, it is common to have a hot-water store and also a cool storage unit. This has the useful side effect of allowing cooling to continue long after dark.

Absorption-cooling equipment is quite bulky, with a typical cooling unit taking up the size of a large cupboard. Given the additional need to pipe cool water or chilled air around the building, the technology is best suited to being installed as a new home is built, rather than being retrofitted in existing homes.

In the coming years we should expect further developments in the field of solar cooling that will bring a virtuous circle of increasing simplicity, lower costs and higher production volumes. At present such systems are much more common on larger industrial buildings, but the compelling vision of an energy-free air-conditioned home is entirely within our grasp.

Mixing solar with other micro-renewables in buildings

There is nothing to stop the designers of a new building, or a retrofitted building, using a mixture of micro-renewables technologies, just as they would use a mix of energy-efficient technologies. Solar photovoltaics and thermal work well together. A tile-shaped frame can contain either solar photovoltaic laminates, or thin-plate solar-thermal collectors, and use of both types of tile provides solar combined-heat-and-power in the roof. But often there are good reasons for mixing other technologies with solar. A building may be partially shaded, for example, limiting the useful area for solar. It may be close to a very attractive source of sustainably harvested wood, making a biomass boiler attractive.

A mix-and-match approach can also make load matching easier.

A biogas micro-combined-heat-and-power unit, for example, can provide heating and electricity in winter, when a solar PV roof is less effective at generating electricity. In summer, when heating is not needed (and electricity is therefore not generated by the biogas unit) solar PV comes into its own. Let's take a quick look at some of the most popular micro-renewables for buildings.

Small- and medium-scale wind

Wind turbines work best in smooth, unobstructed wind, which is why the machines are usually installed on the tops of hills. The urban environment is not ideal in this regard, as buildings introduce turbulence into the wind, reducing the effectiveness of the turbines unless they are installed high up away from trees and other buildings. (The wind 'wake' created by a building can be two or three times its height and extend hundreds of metres downstream.) Another feature of smaller machines is that they rotate at higher speeds, which increases the possibility of noise as well as the problem of light-flicker from the turbine blades. Nevertheless, correctly sited small- and medium-scale wind turbines in and around buildings can make a contribution to onsite or near-site energy for buildings, alongside solar or on their own.

A wind turbine installed at Glasgow Science Centre.
(Image: provenenergyimagelibrary.com)

Air-source and ground-source heat pumps

A heat pump is in effect like a fridge: the low-grade heat inside the fridge is pumped out (to keep the fridge interior cool) and the grill on the back of the fridge gets hot. Working like a reverse refrigerator, ground-sourced heat pumps extract the solar warmth stored a few metres down in the ground. They use a mix of chilled water and antifreeze in a coil of pipe to pump this heat into the home for hot water and heating. They can also be used in reverse, in hot summers, to pump heat out of the building into the ground, so reducing or eliminating the need for air conditioning. Some electricity is needed to work the pump, but of course that can come from other renewable sources such as solar photovoltaics. 1 unit of electricity (a kilowatt-hour) can deliver 2–3 kilowatt-hours of heat.

Biomass

In many rural areas logs are an attractive, sustainable and low-carbon source of heating energy. Other potentially good 'biomass' energy sources include agricultural residues such as bagasse (sugar cane husks) and

forestry thinnings. Fuels like these release CO_2 when they're burned, but only in proportion to the amount they pulled from the atmosphere as they grew. Of course, processing the residues requires some energy, so they're not entirely carbon-neutral, and if the fuels are transported long distances the environmental benefit can be significantly reduced. Nevertheless, local biomass heating can be a very environmentally friendly option in rural areas.

Cost comparisons

The UK government has an ambitious target to make all new homes emit zero carbon by 2016. To set milestones en route to this target it uses something called the Code for Sustainable Homes. This has 6 levels of achievement, with level 6 being the zero-carbon objective. A team of consultants recently analysed for the government which combination of microgeneration technologies was the cheapest way to meet the Code level 3 standard, which requires a 25% reduction

Biomass: wood pellets for domestic use.

in CO_2 emissions for all new dwellings built from 2010. The combination that proved cheapest, for every housing category (i.e. flats and houses in city infill, market-town and urban-regeneration developments) was best-practice energy efficiency married with solar photovoltaics. This combination was found to be less expensive than advanced energy-efficiency measures alone and in most cases much cheaper than the renewable-heat alternatives, including solar thermal plus energy efficiency.

There is a similar story at Code level 4 standard. Only in urban-regeneration properties is PV marginally more expensive per dwelling than biomass CHP plus energy efficiency, but still it is cheaper than all other options including energy efficiency alone. These findings are also based

on an assumption of 0.43 kilograms of carbon dioxide saved per kilowatt-hour of photovoltaic electricity generated, not the assumption used in current building regulations, which is 0.568 kilograms per kilowatt-hour. Even allowing for that, the basic message is that solar photovoltaics is the cheapest renewable technology at lower levels of the code.[34]

This encouraging result allows advocates of solar energy to envisage what might be possible if solar technologies were accelerated into truly mass markets. That vision is the subject of the next chapter.

Desalination

In many parts of the world population growth is creating greater demand for fresh water and the sea is often the only source. The 1.3 billion cubic kilometres of seawater contained in the world's oceans is not suitable for drinking or agriculture due to the high levels of dissolved salt, but the salt can be removed in a process known as desalination.

The most abundant source of fresh water is rain – or other forms of precipitation. Natural desalination occurs when water evaporates from the sea and clouds of pure water vapour form – leaving the salt behind in the sea. Some of the water in these clouds falls over land and, when gathered in rivers, reservoirs and water butts, gives us a supply of fresh water. Where naturally occurring fresh water is not available then artificial desalination can be used. There are broadly three technologies in use today: solar stills, vacuum distillation (or flash distillation), and reverse osmosis.

Solar stills can be used if only small amounts of water are required. A transparent cover placed over a container of salty water will cause the water to warm up and evaporate, and water to condense on the inside of the transparent cover. As the water condenses it can be collected for us. Solar stills like this can be very simple – using sheets of polythene over small ponds – or more sophisticated greenhouse-type structures. Yields of 5 litres per day per m^2 of pond area are possible in areas with high radiation, with no running costs as the energy is free.

In vacuum distillation, seawater is heated in a container held at low pressure. The low pressure reduces the temperature at which the water boils – thereby reducing the amount of energy needed to create the pure water vapour. The water vapour (steam) can be collected and allowed to condense into a separate container, giving a supply of fresh water. This process is also referred to as flash distillation, as the seawater is pumped through a throttle valve into a low-pressure container where the pure water 'flashes off' or evaporates and can be collected and condensed.

Reverse osmosis uses a membrane which is permeable to water but not salt. Pressurised seawater is in effect squeezed through this membrane leaving the salt on one side and fresh water on the other. Both the vacuum deposition and the reverse osmosis processes use large amounts of energy, but reverse osmosis is much more energy efficient as no water heating or boiling is needed.

The most efficient (reverse osmosis-based) desalination plants consume about 5 kilowatt-hours of energy per cubic metre of fresh water produced.[35] Multiple-stage flash processes can be 4 or 5 times as energy hungry. Despite this the theoretical energy requirements to desalinate water are less than 1 kilowatt-hour per m³ so the technology has clearly some evolution to undergo before it is mature.

The bulk of the world's large-scale desalination plants are located in the Middle East, the largest being the Jebel Ali desalination plant in the United Arab Emirates. This plant uses the multistage flash distillation and can produce 300 million m³ per annum.

Solar photovoltaic facade in
Manchester, UK.
(Image: Daniel Hopkinson)

Chapter 5
THE SOLAR REVOLUTION

Visions of Solar in Action
Today and Tomorrow

Anyone formulating a strategy needs to begin with the end in mind. That being the case, let us describe a set of visions of how solar energy could work in a sustainable global energy mix some years from now. In each case, we'll also survey the state of play today, to assess the gaps that have to be closed.

Vision 1: Carbon armies rebuilding economies after the great financial crisis

Any discussion of energy strategies or policies has to begin with improving energy efficiency. Currently a huge amount of the world's energy is frittered away by leaky homes, gas-guzzling engines and inefficient appliances. A mountain of options for boosting efficiency exists around the world, and the mining of this mountain would involve countless short-payback investments; millions of new jobs; gigatonnes of carbon savings; easy, quick income generation – national and individual; and less capital expenditure needed for generation. In a world entering a full-blown economic recession, job creation and quick fiscal savings become particularly vital.

Solar installers at work
in the USA.
(Image: Concentrix)

97

The vision

The energy-efficiency mountain is well on its way to being fully mined, millions of new jobs are being created in a green version of the new deal pursued in the depression of the 1930s, and hundreds of billions of dollars of savings are being made in reduced energy costs. Crucially, the challenge faced by solar and other renewables in powering the world is being essentially halved: far less renewable power is needed than would have been required in a business-as-usual world with steadily growing energy use (3.5% per year for the years 2005–2007).

The jobs being created in improving efficiencies often overlap with the deployment of renewable microgeneration. A major new sector has emerged in the energy industry: the energy services sector. The aim of most energy companies is now to provide heat, light and motive power, not to generate ever more power, or find new oil and gas reserves, or build inefficient engines and all the other manifestations of a supply-dominated mindset.

Some of the energy services companies, or ESCOs, are companies from the old energy sectors that have themselves transitioned rapidly towards demand management and clean-tech manufacturing. They are manufacturing and installing energy-efficient technologies as well as the full spectrum of renewable-energy technologies. They are supported by a plethora of companies involved in producing and implementing smart-grid technologies and energy-storage systems.

The energy services sector has proved to be much more labour-intensive than the old energy industries, and is now dominated by jobs in low-carbon services and manufacturing. The number of jobs in the traditional energy sectors is fast dwindling as the world switches course at a pace that has taken many people by surprise. (Which is just as well, when you consider the impossible aged workforce of the old oil and nuclear industries.)

Vision 1: Potential

People and buildings do not have an innate energy demand; they have a need for heat, light and mobility. In the case of electricity demand, A-rated appliances, devices with low-energy standby modes, and of course the off switch, can all help reduce demand. Passive solar design can maximise the use of daylight, thereby reducing the need for electric lighting. Motion and light-level sensors can reduce the time lights are needlessly on, or are over-bright, without the need for the building occupants consciously to switch off the lights.

Natural heating and ventilation can further reduce the need for energy services. But improving insulation is even more important for existing homes. In the UK injecting insulating foam between the double layers that make up most external walls built since 1930 takes a contractor just a few hours but permanently reduces heating-energy consumption – and bills – by around 15%. Solid external walls can also be insulated by applying decorative weatherproofing, and internal walls by applying insulation boards or infilled wooden battens. Almost a third of heat-loss in the typical building is through the roof. Adding a 27 centimetre layer of insulation to the loft is the simplest and cheapest of all the efficiency measures, reducing the typical UK heating bill by 15%. A further fifth of lost heat escapes from poorly insulated window frames and single-glazed windows, so double-glazing can cut heating bills further. The same principle applies, on a smaller scale, to all the other ways heat can escape. Doors, windows and floorboards can all be simply draught-proofed with sealant.

Loft insulation installer in the UK.
(Image: nationalinsulation association.org.uk)

The McKinsey Global Institute (MGI), the research arm of the well-known global consultancy, believes the world could more than halve projected energy-demand growth, by using existing technology profitably. The investments needed to do this would earn an average return of 17%, and a minimum of 10%. These calculations cover all sectors of the global economy, but the sector with the most reduction potential is the residential sector, which offers 24% of the potential for improving energy productivity. Around $170 billion would have to be spent by 2020 to hit the McKinsey target, but the returns would be quick, and anyway that figure is a mere 1.6% of today's global annual investment in fixed capital. No wonder energy-efficiency pundits refer to its savings potential as 'negawatts'.

So what is holding the world back?

One answer, bizarrely, is that for a long time electricity and fuel have been too cheap to warrant the up-front expenditure on energy-saving improvements to buildings. Another is that landlords have had no incentive to invest in efficiency on behalf of their tenants. A third is that many governments have ignored the potential for leadership in energy efficiency – in their own buildings as well as the wider housing stock. With the current high energy prices, and the many forecasts that prices can only go up in the long run when it comes to traditional energy supply, the situation is surely set to change.

Vision 1: Progress

Business is booming for energy services companies that help families and organisations reduce their energy bills. In America, ESCO income has grown from 3% a year in the early years of the century to 22% in 2006. An ESCO usually does an audit of the client's buildings, designs an energy-reduction scheme, borrows money to pay for the energy-saving equipment, and makes its return on the money saved, or a part thereof. The client pays nothing, only saves.[1]

A recent social experiment in the UK gives a flavour of the potential for ESCOs in the 'green new deal'. British Gas encouraged eight typical British streets to compete with each other to cut energy bills. Advisors helped householders pick the low-hanging energy-savings fruit, which they did with ease, quickly saving 30% on energy bills and 20% on greenhouse emissions. The Institute for Public Policy Research refereed the exercise. They calculated what would be needed to replicate it nationwide. The answer proved to be 10,000 energy-auditor jobs at an outlay of half a billion pounds. This outlay, for these 10,000 green jobs – a division in the British carbon army – would save a massive £4.6 billion in the first year, and more than that in every subsequent year.

Solar PV tile installation in the UK.

That has to be described as a good return on investment.[2]

Germany has already done this kind of thing on an enviable scale. Between 2001 and 2006, in an earlier recession in the building industry, the government invested the euro equivalent of $5.2 billion in retrofitting German apartments with efficient-energy technologies. This investment leveraged a further $19 billion of private money and created 140,000 jobs. The government recouped $4 billion of the original investment in tax paid by the new workers and unemployment benefits avoided. This, again, is very impressive payback even before you consider the energy and carbon savings.[3]

Nations need to generate power as well as save energy, of course. Here the good news, as we saw on p.85, is that renewables create more jobs per unit of power and per pound invested than conventional generation. (Public transport, we should also note, similarly generates more jobs than vehicle manufacture and use.) Out of at least 2.3 million people employed in the global renewables industry as of 2006, some 794,000 were employed in the solar industry (about 624,000 in solar thermal and about 170,000 in PV, 40,000 of them in Germany). Yet the global renewables industry is minuscule compared to its potential. A global renewables mobilisation could create many millions of jobs at a pace that would amaze most people. In Germany, 25,000 new solar PV jobs were created between 2004 and 2007, and total renewables-industry employment grew to 260,000 in less than ten years. These are local jobs, interesting, healthy jobs, and generally jobs that do not end up outsourced overseas.

PV installation in Germany. (Image: SMA Solar Technology AG)

It is not as though the world is starting from square one in achieving the first vision. The global market for environmental goods and services is worth £1,300 billion per year ($1.3 trillion), half of which is in energy efficiency, according to the UN.[4] At the time of writing, politicians on both sides of the Atlantic are talking about the need for a green new deal, while the International Energy Agency has called for a clean-energy new deal. In industry, Deutsche Bank has produced a report calling for the creation of up to 25 million 'green' jobs. This document, which seems more typical of a progressive think tank than of one of the world's biggest banks, proposes the creation of National Infrastructure Banks to fund the commercialisation and roll-out of green opportunities, and public–private partnerships to scale up these initiatives. The spending programme should focus on a 'green sweet spot' including energy efficiency in buildings, the electric power grid, renewable power and public transport. The report also suggests that a National Infrastructure Bank 'could underwrite state, local and private-sector bonds, potentially enabling public–private partnerships or unlocking other appropriate financing for private-sector projects'.[5]

Surely, with support like this, we have to conclude that Vision 1 is not only feasible, but overdue.

Solar installation on a large scale, Fontana, California, USA.
(Image: First Solar, Edison)

Vision 2: Most buildings, old and new, transforming into solar power plants

Winning the race to cut greenhouse-gas emissions and wean ourselves off fossil-fuel dependence will require action across all economic sectors. But one sector is particularly important, and particularly suitable for the application of solar. Energy use in buildings accounts for more than half of carbon dioxide emissions in a country like the UK. The emissions come about as a result of combustion of oil and gas inside the building to produce heating and hot water, and from the use of electricity, which is typically generated by remote power stations and transmitted to the buildings over the national grid and local electricity-distribution networks.

The vision

Solar technologies are being deployed routinely: most new buildings and many existing buildings have all or part of their electricity, and/or heating, and/or cooling provided by solar. On many roofs, solar-photovoltaic tiles and modules provide electricity, and solar-thermal tiles and collectors provide hot water, and energy for cooling. Other elements of buildings are solarised. Façades are covered in solar-photovoltaic modules and/or thermal collectors. Flexible thin-film photovoltaics are routinely bonded to thin metal, and membranes, and so curved roofs or modern tent-like architectural designs are routinely solarised. Even windows generate solar electricity. Laser-scribed thin-film glass laminates trap light for electricity generation while letting a restful amount of light through, leaving the outside world visible to occupants within.

In many buildings, solar is not used alone. The full family of micropower technologies – renewable and gas-powered – are deployed in strategic harness across the stock of residential and commercial properties. Some buildings have three or more members of the family working at once. So much electricity is being generated in some buildings that they are essentially mini power plants: exporting into the smart grid, charging storage devices, or both.

The energy-efficiency potential of Vision 1 greatly amplifies the potential of Vision 2. In the smart grid, 'smart buildings' that control when they use power, or store energy when electricity is cheap, avoid the need to buy expensive power at times of high demand while also spreading out demand and preventing the need for the oldest power stations to be fired up to meet peaks in demand.

Vision 2: Potential

A typical British home employing extensive energy efficiency, or a modern new build, can consume less than as 2,000 kilowatt-hours of electricity a year. For example, a 62 m² new-build end-of-terrace property built to level 4 of the Code for Sustainable Homes will consume 1,993 kilowatt-hours of electricity, according to the UK government's Standard Assessment Procedure. A 2.4 kilowatt PV system, requiring as few as 15 m² of roof space, would produce 2,040 kilowatt-hours a year, allowing excess electricity to be sold to the utility company. In the future, receiving income from electricity sales is going to be something enjoyed by growing numbers of people. For this reason, and because normal electricity prices are sure to rise, adding solar at home will become a useful way for people to top up their pension.[6]

The potential is huge, even in a cloudy country like the UK. The absolute resource potential for solar PV on British buildings (i.e. if PV was put on all available roofs and walls) is fully 460 terawatt hours per year, which is more than the entire national electricity demand today of around 400 terawatt hours per year. Ground-mounted PV would add to this considerably more. Of course, it wouldn't be sensible to aim for this absolute potential, when there are so many other renewable energy technologies, and when we have such huge energy savings yet to make. But even a small fraction of the absolute potential would make a big difference. Using just south facing roofs and facades, for example, would generate 140 terawatt hours per year.[7]

There is also huge scope for solar on non-domestic buildings in the UK. A recent report for the UK government calculates that 33 gigawatts could be installed with the potential to provide more than 25 terawatt-hours per year on 1.5 million individual sites, if current economics were not a constraint. At a feed-in tariff of 40p per kilowatt-hour, the potential of all the buildings in the non-domestic sector could be unlocked by 2014. With installation at a constant 5% per annum, 15.4 gigawatts could be installed by 2020. That would provide 12.5 terawatt-hours a year of electricity, more than 10% of the UK renewable-electricity target, saving 5.4 million tonnes of CO_2 per year.[8] In the real as opposed to the calculated world, beyond grid parity and with the advent of a mass market, the rate of installation would of course accelerate.

Vision 2: Progress

The UK's first solar roof-tile home was a humble two-bedroom terrace in Richmond with a small 1.6 kilowatt photovoltaic installation. In the first year of its life, the roof generated over 1,100 kilowatt-hours of electricity. Using the most efficient lights and appliances available in the shops, the occupants (the editor of this book, his daughter and her husband) were able to keep the consumption down to just over 1,000 kilowatt-hours. This made the small home a net exporter of electricity, by some 14%: a mini-power station, in other words. That was in 2000. Now in cloudy Britain there are hundreds of solar-electric roofs.

In equally cloudy Germany – which unlike Britain has enjoyed a feed-in tariff to guarantee a fixed price for electricity fed back into the grid – there are hundreds of thousands. Ready-to-occupy solar houses are being provided at normal market prices. For example, the first turnkey affordable house is the 'Energetikhaus 100', occupied since 2006, which covered 97% of its heating demand from solar hot water during its first winter. The living space is 137 square metres and the south roof includes 69 square metres of solar collectors. Another design, the Kroiss 'plus-energy' house in Austria, is built of wood and insulated to the famously high 'passivhaus' standard. On the roof, 10.35 kilowatts of solar photovoltaics provide more electricity than the occupants (two adults and two children) need, and 17.4 square metres of solar-thermal collectors provide all the hot water required. [9]

The UK's, and the Editor's, first solar roof, London.

As for Europe's first solar street, perhaps surprisingly it is in Yorkshire, England, where the forward-thinking South Yorkshire Housing Association has installed solar PV and solar-thermal roof tiles on each of twenty-three new, affordable, stylish and low-carbon homes. As a result, buyers and tenants alike benefit from hot water, electricity and protection from energy-price rises. To the observer, the tiles look exactly like standard roof tiles.

Solar technology has been combined with state-of-the-art insulation to minimise the need for central heating except on the coldest days. The solar hot-water tiles generate over 60% of the hot-water requirements of the three-bedroom home while the solar-electric tiles generate around 2,500 kilowatt-hours of electricity per year. Any excess electricity generated is simply exported to the grid via an export electricity meter. If the UK had a domestic feed-in tariff like France's, payments for exported electricity would make a useful income for the household.

Traditional roofs contrasted with the modern alternative. Solar tiles in Rotherham, Yorkshire, UK.

Progress is not limited to the West. China installed more than 18 gigawatts of solar-thermal collectors in 2006, adding substantially to the total installed capacity of 128 gigawatts worldwide. China accounted for almost 70% of all solar-thermal collector capacity installed in the world that year, up from 45% in 2005. Yet in per capita terms, China lags far behind the leading nation, Cyprus, where there are 680 kilowatts of solar thermal per 1,000 inhabitants.[10] Imagine if China, currently at less than 50 kilowatts per 1,000 people, had a Cyprus-level per-capita performance. It could, easily.

Opposite: The solar PV clad CFS tower in Manchester, UK.

Solarisation of commercial and industrial buildings is also slowly making progress. In rainy Manchester, for example, marble-like blue polycrystalline solar PV modules clad all three sides of the service core of the T-shaped 28-storey headquarters of a financial services group. The CFS tower is the tallest building in northern England, a national landmark protected as a listed building. More than 7,000 panels cover almost 4,000 square metres of the building, giving a peak power yield of 391 kilowatts.

Why did CFS choose to use solar, in such a cloudy place? There are two main reasons. First, the original core of the building was previously clad with 14 million small grey ornamental ceramic tiles. These started to fall off within six months, and so became a hazard. The façade had to be replaced. Why not replace the beautiful but functionless old façade with a beautiful but functional solar one, CFS reasoned. The price would be roughly the same, and the new façade would give free electricity for decades. Secondly, CFS is a company that believes in doing all it can to promote a sustainable future. By solarising its headquarters, it could show that solar works well enough in cloudy countries, and send out a message that, if enough companies and individuals follow in its footsteps, solar PV can reach a level of demand where the cost of manufacturing falls far enough to cross the rising price of traditional energy. The 180,000 kilowatt-hours-plus of free electricity generated each year saves more than 100 tonnes of carbon dioxide a year.

Flat roofs are being used for even larger carbon savings. Giant American retailers Wal-Mart, Kohl's, Safeway and others are promising to put solar on almost every big store if the government goes ahead with tax rebates for solar. If Wal-Mart were to cover every Sam's Club and Wal-Mart store, 23 square miles of solar panels would be needed: an area the size of Manhattan.[11]

Were Wal-Mart a nation instead of a corporation, it would have the eighteenth largest economy in the world. With the likes of this giant becoming engaged in solarisation, and the evidence from progress to date, we surely have to conclude that Vision 2 is feasible.

Vision 3: Glittering fields of steel and silicon generating power for cities far away

In the future world, the major energy utilities will have broken with the past, not just in terms of the provision of customer services but in terms of generation, too.

The vision

The power companies have been much influenced by governments at federal and state level, who require them to have 'renewable portfolio standards', i.e. a specified amount of renewable power – sometimes specifically solar – in their generating mix by a certain year. As a result of this, plus increasingly vocal customer and shareholder demands, and of course increasingly enlightened business leadership, the utilities in the sunny countries are installing large solar-thermal and solar-photovoltaic electric power plants, rather than coal and gas ones.

Beyond this, in a new spirit of international cooperation imposed by the global financial crisis, nations are coming together to pool their power requirements more than was the case before. Massive solar power plants are under construction in sunny nations where only a fraction of the power generated will be needed domestically. The rest will be exported along transnational grids to other countries.

Vision 3: Potential

We described the potential for utility-scale solar-thermal and photovoltaic power plants in the south-western United States on p52. In California, Pacific Gas and Electric seems to appreciate the potential, and plans to install 800 megawatts of solar PV in two giant solar farms by 2013. This 12 square mile development will dwarf today's largest American solar PV farm (a mere 14 megawatts on the Nellis Air Force base) and produce more power than the total PV capacity currently connected to the US grid (473 megawatts). According to the current plan, as we saw on p.74, it will generate 1.65 billion kilowatt-hours a year, peaking in the afternoon on the sunniest days when the region's electricity demand is at its highest.[12] As PV manufacturing costs continue to fall, there is increasingly little to stop other utilities following Pacific Gas and Electric's lead, or the latter from moving from their first utility-scale solar farm to many.

The European Commission's Institute for Energy is just one organisation that envisages many 50–200 megawatt solar plants in Africa, linked by a high-voltage DC transmission grid to Europe. Such a grid loses less energy over distance than AC: 3% per 1,000 kilometres. To provide all Europe's electricity needs this way would require, in principle, the capture of just 0.3% of the light falling on the Sahara Desert and the Middle Eastern deserts. This could be done in an area smaller than the equivalent of Wales, covered either with concentrated solar power or photovoltaics, generating up to three times more electricity than equivalent power plants could in Northern Europe. The grid could also capture wind from across Europe and North Africa. According to European Commission estimates, by 2050 North Africa could be providing 100 gigawatts to Europe, at a total investment of €450 billion (c. €10 billion a year). In context, this is not too expensive: the IEA now calculates that the world needs to spend $45 trillion (£22.5 trillion) on energy systems over the next thirty years.[13]

Vision 3: Progress

Algeria is already working on a combined solar and gas (CSP) plant aiming to export 6 gigawatts of electricity to Europe by 2020. So far, however, most of the progress has been achieved within Europe and the US.

More than fifty CSP projects have already been approved around Spain. If all these are built, they will be capable of generating 2 gigawatts by 2015. As the projects get bigger, the price falls, and analysts expect grid parity within a decade. For example, near Seville the 11 megawatt PS10 plant, a power tower project, has been active for two years now, and the PS20 power tower is on course for completion. More than 1,000 mirrors, each half the size of a tennis court, are being positioned to focus the sun's rays on the tower. They will generate 20 megawatts of electricity, enough for 11,000 homes, at a cost of €80 million. The next project for Abengoa, the Spanish energy company behind the plant, is a 50 megawatt version that would give electricity in the night as well as the day. This will be possible because 50% of the electricity produced during daylight will be used to heat molten salt. Energy is released from the salt at night to drive the turbine. Tests so far show up to 8 hours of electricity can be stored by heating 28,000 tonnes of salt, in tanks, to more than 220° Celsius.[14]

Meanwhile, the first commercial parabolic-trough solar-thermal plant in Europe has begun operation in Andalusia. This is one of what

Opposite:
Solar powering the future in Abu Dhabi.
In the background is the construction site of Masdar, the world's first solar city.
(Image: Environmena)

The PS10 solar-thermal power tower installation in Sanlúcar la Mayor, Seville, Spain.
(Image: Abengoa Solar)

will eventually be three Andasol plants south of Grenada. In each plant, 2 million km² of reflectors direct the sun's energy onto 2,000 cubic metres of heat-exchanger fluid, achieving temperatures of 400° Celsius. So that the plant can operate at night, two salt reservoirs are employed, containing 28,500 tonnes of molten salt in all. A 'cold' container holds salt heated to 292° Celsius, while the hot one reaches 386° Celsius. These lock up enough heat to run steam turbines for more than 7 hours. Each of the three plants has a capacity of 50 megawatts, is capable of supplying up to 200,000 people, and has construction costs of some €300 million. At the first plant, the salt-storage system began operating in November 2008. The second is under construction and the third is being planned.[15]

In the south-western United States, as a result of the growing interest of Pacific Gas and Electric and other energy companies, a land grab is under way for solar sites. In the Mojave Desert 104 claims have been received for nearly a million acres of land. That's enough to provide a theoretical 60 gigawatts of electricity – almost double California's total existing capacity. Some companies have paid more than $10,000 an acre for scrubland. One player, Solar Investments – a subsidiary of Goldman Sachs – will probably either partner with developers or sell leases. The bank has requested permission from the Bureau of Land Management to install and monitor a range of thermal and photovoltaic technologies.[16]

Some of these plans will not come to fruition, no doubt. Some of the early giant plants may hit technical teething problems. But the momentum is such that failed plans will surely be replaced by other plans yet to be revealed. Technical problems will probably be overcome, as they have been on so many other technological frontiers. After all, utility-scale solar plants hardly involve rocket science. Vision 3 surely has to be deemed feasible.

Vision 4: Solar photovoltaics at grid parity everywhere, years earlier than most expected

The terms by which the costs of solar energy and traditional energy are compared are generally skewed in favour of the latter. For solar advocates, this seems unfair. Why should energy economists be allowed to place no value on solar as a construction or design material, we ask? Why should they be so fixated on a snapshot of today's energy pricing when solar prices are heading inevitably downward, and conventional energy prices are heading inevitably upward? Even in this narrow frame of reference, however, solar is poised for a breakthrough in the form of grid parity.

The vision

Grid parity – where the price of solar-photovoltaic electricity is equivalent to or less than the price of grid electricity – has arrived across almost every part of the world where electricity grids are operated, including in the cloudy temperate countries.

Vision 4: Potential

Grid parity, where the cost per unit generated of investing in solar draws level with that of grid electricity, is the point at which we can expect to see mass solar markets take off.* In June 2008, McKinsey estimated that solar generation will be at grid parity in at least ten markets around the world by 2020.[17] The European Photovoltaic Industry Association (EPIA) estimates grid parity in most European country markets, including Northern Europe, by the same year.

Some others think differently. For example, the Intergovernmental Panel on Climate Change, an expert body whose thinking is very influential among governments, does not expect to see cost-competitive

---

* Note that grid parity is defined here (and in other studies) as the point at which the unit price of solar-photovoltaic electricity draws level with the price of grid electricity, as a single snapshot in time without discounting future generation. Those investing in PV for purely financial reasons will require an appropriate return on investment, and therefore may only start to invest heavily when the PV price incorporates this return – i.e. slightly after grid parity itself.

solar-photovoltaic electricity until after 2030. The UK government's Climate Change Committee thinks that the technology will not be competitive until much later still. Solar PV costs are 30–35 cents per kilowatt-hour in 'areas of good irradiation', the committee has asserted, arguing that it will take 'several decades' if not until the 'late twenty-first century' for PV to be cost competitive anywhere, and perhaps never in the UK.

Vision 4: Progress

The reality is that in some markets grid parity for crystalline solar PV is nearing, and it has already arrived for thin-film solar PV. This is before external environmental costs (such as carbon pollution) are factored in to the costs. Figure 20 shows figures compiled by investment bank Lazard, in 2008 dollars.[18] Costs for gas range from 13.3 cents per kilowatt-hour in a type of modern gas power plant (IGCC: Integrated Gas Combined Cycle) with 90% carbon capture, down to 7.3 cents per kilowatt-hour for more straightforward gas plants. When gas is used for peaking power it can be as high as 33.4 cents per kilowatt-hour. Coal ranges between 13.5 and 7.4 cents per kilowatt-hour.

Crystalline silicon costs, according to Lazard, range from 16.8 cents per kilowatt-hour for fixed systems to as low as 11.8 cents per kilowatt-hour for single-axis tracking systems. Crystalline cell makers forecast energy costs as low as 9 cents per kilowatt-hour by 2010, based on economies of scale associated with projected production growth and increasing efficiencies (at a fully installed system cost of around $4,000 per kilowatt).

Thin-film generating costs today range from 12.4 to 9 cents per kilowatt-hour. Based on projections by First Solar, the leading manufacturer, total systems costs in 2012 can be as low as $2,000 per kilowatt, which would give a cost of energy of 6.2 cents per kilowatt-hour. All these figures for PV include the US production tax credit.

Solar thermal costs range between 14.5 cents per kilowatt-hour and 9 cents per kilowatt-hour.

These figures, it should be emphasised, are snapshots frozen in time. Costs for gas and coal are on a clear upward trend over time. In the US, for example, rates for just siting building and operating plants have risen an average of 3.5% per year for the past six decades. The costs of solar, meanwhile, are coming down.

Levelized Cost of Energy Comparison (Lazard)

Figure 20: Levelised energy
costs as estimated by
investment bank Lazard.[18]

Moreover, comparing the cost of solar PV electricity generation with that of generation from centralised fossil-fuel plants is somewhat misleading. Since solar PV is a form of decentralised energy available at the point of use, the appropriate comparison is with the retail and commercial rates actually paid by customers for grid-supplied electricity. After all, from the perspective of a householder, each unit of PV electricity consumed is a unit of retail-price grid electricity avoided.

In the UK, for example, domestic-retail electricity prices reached 14p per kilowatt-hour in 2008. At the time, a typical UK householder with a solar roof was paying approximately 23p per unit for their PV-generated electricity. But this gap is closing rapidly as retail electricity prices rise (UK electricity-price rises have averaged 5% per year over the last ten years) and PV costs fall. Figure 21 shows that grid parity for domestic PV customers could be reached as early as 2013, assuming modest increases in grid electricity inflation and significant cost reductions (as widely predicted) for solar technology. Residential grid parity, therefore, could be

close even in cloudy Britain. Even for commercial, non-domestic customers in the UK, parity could arrive before 2020. As Figure 22 shows, the lower costs possible in a sunny climate like southern California bring grid parity closer still. And California is also a market where real-time pricing of electricity means that PV's predictable daytime output, matching patterns of consumption, improves its cost relative to conventional grid supply still further.

The 2006 Stern Review on the economics of climate change emphasised that solar PV will achieve considerable economies in production, thanks to technical developments in manufacture, rapid feedback of experience and short lead times for investment (compared to the 3–6 years or longer needed for 'conventional' generating plant). Even the IEA – an organisation not known for its sympathetic view of renewables – predicted in its 'World Energy Outlook 2008' that in China, Europe and the US, solar PV prices will fall significantly to $3,600–3,900 per kilowatt installed by 2015 and to $2,450–2,750 by 2030.

While solar manufacturing costs fell as usual between 2006

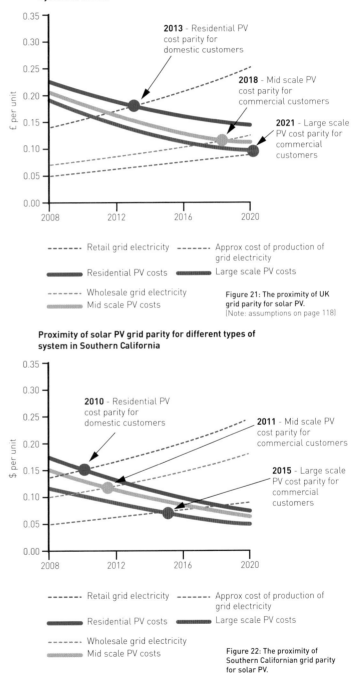

Figure 21: The proximity of UK grid parity for solar PV.
(Note: assumptions on page 118)

Figure 22: The proximity of Southern Californian grid parity for solar PV.

and 2008, installed PV prices actually rose. This anomaly was largely the result of under-capacity in the industry, particularly in the area of silicon processing. A boom in silicon-production facilities around the world, with more than seventy facilities being constructed or planned as of the end of 2007, has resolved the problem. The global economic downturn, of course, has caused prices to fall even faster.

Grid parity assumptions.

Input		Base scenario (conservative)
PV 2008	large	£4.5
Installed prices £/Wp:	mid	£4.8
(includes lifetime maintenance)	small	£5.3 (BERR Renewable Energy Strategy, after Element Energy 2008; also LCBP2 figures)
PV price declines		6% p.a. decline (BERR Renewable Energy Strategy, after Element Energy 2008)
PV lifespan / performance		30 years 0.6% p.a. module output decline 850 kWh/kWP (EPIA, PV-GIS, module manufacturer data)
Grid electricity prices:	cost;	5p/unit
	w'sale:	8p/unit
(2008 baseline)	retail:	14p/unit (BERR Quarterly Energy Prices, 2008 median)
Grid electricity inflation		5% (BERR, rises over last 10 years averaged; equivalent to 3% p.a. plus one off 30% spike)

Vision 5: Energy storage plants throughout the smart grid ... many of them also vehicles

Another complaint advocates of renewable microgeneration have about price per kilowatt-hour as the benchmark in energy policy is that no credit is given for the ease of turning supply on and off. In a world with smart grids, this is going to be very important.

The vision

Most vehicles are battery-powered, and the electricity used to charge them is supplied by renewables such as solar and wind. Cars are able to travel long distances between charges – not that they particularly need to, as recharge and battery-swapping stations are beginning to be almost as common as petrol-filling stations once were. When parked, hybrid and electric vehicles are plugged in.

Well-charged vehicles can discharge electricity into homes and commercial premises when necessary, helping remove the problems of intermittent energy sources such as wind, and earning income for the car owner. All the electricity-flows are tuned and optimised on smart grids to maximise efficiency and energy security.

Vision 5: Potential

Car companies increasingly are betting on electricity as the transport fuel of the future, and there is no reason why this electricity cannot be provided by renewables. Electric vehicles powered by lithium-ion batteries are poised to go mainstream. This is in major part because breakthroughs in this particular member of the clean-tech family now mean the batteries are light enough and small enough to fit into cars without weighing them down. Tighter vehicle-emissions regulations are also helping. In addition, as so often happens, higher manufacturing volumes are reducing the price differential between the disruptive technology (electric cars) and the traditional alternative (hybrid and petrol cars).[19] Renault-Nissan says it aims to lead the industry in all-electric cars. They are teaming up with Project Better Place to take electric vehicles to California, Hawaii, Israel, Denmark and Portugal. By 2012, they intend to have a range of electric

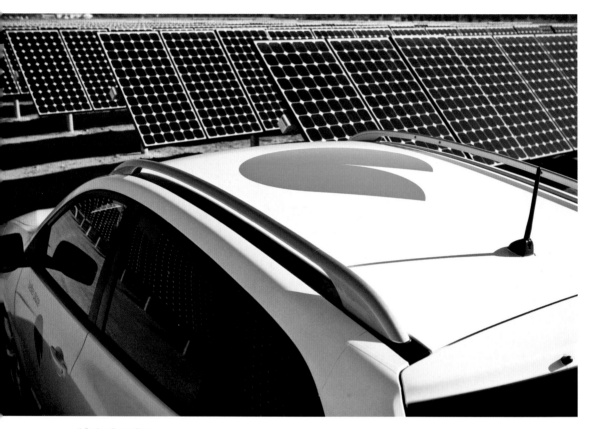

A Project Better Place electric car at a Californian solar farm.
(Image: www.betterplace.com)

vehicles (EVs) in all main markets offered at prices lower than equivalent petrol models. Nissan and NEC are investing heavily in the lithium-ion batteries needed to make this happen.[20]

Vision 5: Progress

Viewing progress with electric cars through the prism of sports car performance will seem counterintuitive (or worse) to some people. But in a sense, sports cars are ideally placed to convey the fact that a solar-electric world can do most of the things a petrol-powered world can, and a lot more besides.

First unveiled in the US in 2006, and first driven in the UK in May 2008, the Tesla Roadster shows the potential of electric vehicles by putting a high-performance battery in a high-performance sports car. The 'Green Rocket', as the *Daily Mail* called the car, is faster than all but a few petrol-powered cars. It accelerates from zero to 60 mph in less than four seconds, has a top speed of 150 mph, and a range of 250 miles.[21] The Tesla overcomes objections raised about earlier battery cars, like the GM EV1 of the 1990s. The best EV1 had nickel-metal hydride batteries, whereas the Tesla uses lithium-ion batteries, which avoid memory problems

The Tesla Roadster.
Image: (www.teslamotors.com)

and have the benefit of being recyclable. Regenerative braking adds to the car's efficiency. Compared to a popular hybrid car the Tesla is twice as efficient and emits half as much CO_2 per mile (assuming it's charged with electricity from a gas power plant). Compared to a typical sports car, it's six times as efficient and emits a tenth of the carbon dioxide.

Of course, if you use renewable electricity, the emissions would be zero. The car needs 0.38 kilowatt-hours of power for each mile of driving, so if you drive it 10,000 miles a year, a 5 kilowatt solar PV array would provide all the electricity needed. An array that size could fit on the roof of many a home. A mere 3.5 hour charge is needed to achieve the 220 mile range, and Tesla says the charge time will shrink to that of a long lunch as their technology improves.[22] Such vehicles are not only being designed by American companies. The first British green sports car was unveiled in 2008. Accelerating from 0–60 in four seconds and travelling more than 200 miles on a charge – just like the Tesla – the Lightning GT even has space for suitcases as well.

Japanese drivers will be the first in the world to be offered plug-in cars by the major car makers: in 2009 by Mitsubishi Motors and Subaru and 2010 by Toyota and Renault-Nissan. Japan plans to build hundreds of quick-recharge stations. Tokyo Electric Power (TEPCO) says it has developed a device that recharges enough of the battery in five minutes to allow a 40 kilometre drive, and ten minutes to allow a 60 kilometre range. The device costs $36,500 and will be installed in supermarkets and other public places. The government, aiming for half of all new car sales to be electric by 2020, is doing its bit: offering discounts to EV drivers on parking, loans, insurance and other things.[23]

The San Francisco Bay Area plans to be an electric-car hot-spot, with local cities, supported by state government, planning to unite and spend

The Project Better Place eRogue, developed in conjunction with Renault-Nissan.
(Image: www.betterplace.com)

as much as $1 billion on charging and battery-swap points. On the other side of the world, Israel has announced a nationwide electric-car project aiming to remove the need for oil imports within a decade. A private plan, with the backing of the President, involves installing 500,000 recharging points and battery-swap stations for electric cars by 2009, with the aim of halving oil dependence within a few years. Solar-electric plants will be built to offset the rest of the oil imports. Project Better Place has raised $200 million for the initial stages of this visionary scheme. The rest of the infrastructure and vehicles are expected to cost a further $800 million. Shai Agassi, the founder, calculates that, if Israel's fleet of 2 million cars were all electric, they would require 2,000 megawatts of electricity per year, entailing an investment of $5 billion in solar plants. This is clearly doable, he believes. He likens the scheme to that of the early infrastructure companies who made the widespread use of mobile phones possible.[24]

Trains are also likely to switch ever more from oil-based fuel to electricity in the years ahead. A wide range of rapidly growing renewables markets, with appropriate storage, will find little difficulty in powering the rail networks of the future.

The trend to electrification of transport is developing against a backdrop of structural change under way in American transport as 'reurbanisers' leave the suburbs in the face of economic woes. Americans drove 11 billion fewer miles in March 2008 than in March 2007. Bus and train use was up 10–15%. While sales of bicycles and scooters were soaring, sales of SUVs were slumping and the CEOs of Ford and GM – switching their production lines to fuel-efficient cars and plug-in hybrids – talked of a 'structural' change rather than a cyclical one. All this will help achieve the vision of cars as mobile power and energy-storage plants. However, there is still a long way to go. In 2007 only 2.2% of car sales in the US were hybrids, let alone all-electric vehicles. Even with that caveat in mind, though, Vision 5 certainly looks feasible, and might become reality much earlier than many commentators currently assume.

Vision 6: Solar farms side by side with food farms on arid coastal plains

There will be little point in networks of electric vehicles acting as energy-storage plants if humankind's ability to feed itself continues to be impaired at the rate it has been in recent decades. The energy-infrastructure renaissance has to be accompanied by an agricultural renaissance, and on a global scale. Solar can play a part here too.

The vision

Huge solar markets have arisen in the Arabian Gulf countries. The example they are setting with solar energy is encouraging people on the many arid coastal plains around the world to solarise their own economies. Some of the regions most affected by the proliferating droughts and shrinking water supplies caused by global warming are finding that they can adapt better using solar for desalination as well as for power. As they do so, they are cutting their own emissions of greenhouse gases, creating an important virtuous circle.

Vision 6: Potential

One might imagine that the Arabian Gulf nations would take a rather dim view of solar energy, sitting as they do on the biggest oil reserves in the world. This is not so. Concern is breaking out about the rate at which their exploding economies, fuelled by petrodollars, are sucking up oil and gas supplies. Hence the new regional interest in all parts of the solar industry.

In parallel, designers and inventors are extending the use of solar power to other vital economic sectors. The Sahara Forest Project is an example. In this design, huge greenhouses twinned with concentrating solar power make cropland without wells, plus energy for local economies. Solar farms run seawater evaporators and pump damp, cool air through the giant greenhouses. The water vapour is condensed at the far end of each greenhouse, and used to irrigate crops in the immediate vicinity. Demonstration projects are up and running in Tenerife, Oman, and the

United Arab Emirates. The designers contend that virtually any vegetable can be grown this way. Nutrients come from local seaweed, or the seawater.[25] In a world increasingly struggling to feed itself, the potential of this kind of project is hard to overstate.

Vision 6: Progress

In Abu Dhabi, ground has been broken on the first renewably powered city. Masdar will be home to 40,000 people living in a car-free zero-waste environment, with solar-powered cooling and water desalination. The city will use 80% less water than conventional cities through the use of grey water, recycling and efficient processes to drastically reduce waste. By 2016, 800 post-graduate students will study photovoltaic engineering, desalination, hydrogen power and other sustainability related courses at the Masdar Institute and 1,500 businesses will work on solar and other alternative energy sources within the city.[26]

The compact, low-rise city – designed by world-famous architects Foster and Partners – aims to be an ecotopia: the world's first fully planned and integrated sustainable city. A light-rail system will link Masdar to neighbouring developments and ultimately Abu Dhabi, and driverless pods using a magnetic guidance system will move people within the city.

Above and right, artists impressions of Foster & Partners plans for Masdar.
(Images: fosterandpartners.com)

Every surface will collect energy. Natural air-cooling techniques are to temper the air within the streets. 'If Masdar does not uplift the spirits,' says Foster, 'then it's not fulfilling its function. Solar can be beautiful. Green is cool. It has to be a better world. You have to have a belief in the future.'

200–230 megawatts of peak power will be needed to run the city-wide systems and 45% of this will come from solar photovoltaics. A 10 megawatt solar PV farm is about to be connected to the grid at the time of writing, supplying 17,500 megawatt-hours a year.[27]

Masdar means 'source' in Arabic. Set in a region of energy-profligate high rises and development without much attention to sustainability or even central planning, Masdar can be seen as a source of hope: a sign that things can be done differently – especially if you build them around solar.[28] Since construction started, Abu Dhabi has become the

first Middle Eastern country to set a renewables target: 7% of power by 2020. Abu Dhabi has also recently tightened its building regulations to cut emissions. Officials at the World Future Energy Summit in 2009 said that the vast majority and maybe all of it will derive from solar. This being the case, it is easy to imagine solar-powered desalination – with the water allocated to civil projects including food cultivation – soon making an appearance on the Gulf Coast, proving the feasibility of Vision 6.

Vision 7: The rural poor reading at night

People in the developed world, and parts of the developing world, have come to take electricity and its many benefits for granted. In other regions, almost 2 billion people still have to get by without access to grid electricity. Governments in these areas do not have the huge sums of

money needed to set up and build and run a centralised power system, and even when they indebt themselves to do so, the grids tend to be unreliable – not to mention a disaster for air quality in the areas around the power stations.

Even where renewables are used in large centralised schemes, the poor often don't benefit. At the time of writing, African governments and banks are discussing an £80 billion hydroelectric power plant scheme on the Congo River. The Grand Inga Dam would potentially double Africa's electricity supply, generating twice the amount of the controversial Three Gorges Dam in China, currently the world's largest. The idea is an old one, and is being resurrected because good financial returns are available based on the availability of carbon credits under the Kyoto Protocol to combat climate change. Critics say it is a white elephant that would leave Congo with mountainous debts. 94% of the people in Congo are without electricity, along with two thirds of all Africans. They would still have little or none because the power lines from the Grand Inga would head to existing industry centres, especially in

A kerosene lantern, converted to clean solar powered LEDS, provides light for children to read by in Muhuru Bay, Kenya.
(Image: SolarAid - Andy Bodycombe)

South Africa. Congo has exported electricity for years from two smaller schemes at Inga, and villagers right next door receive none of it.[29] Here, and elsewhere, people literally lack the energy to lift themselves out of poverty. Micropower solutions – and especially solar solutions – can often make a real difference in combating this inequitable state of affairs.

Let us consider, in this context, the simple kerosene lantern: the most popular means of generating light at night across the developing world. Hundreds of millions of households have little alternative but to use kerosene lanterns, and well over a billion people live on less than a dollar a day.

The vision

Almost every kerosene lantern in the developing world has been converted into a solar lantern, or been replaced with a new solar lantern. The solar-adaptation kit and the new solar lanterns are both far more economic than the older option, not to mention the health benefits, so significant proportions of household incomes are saved. Local people are taught to market the products, and provide after-sales service. The additional incomes involved, plus the money saved on kerosene bills, provide cash to buy food, plus seed and fertiliser, a safety net against hunger, and a first step out of poverty. In this way, another virtuous circle is created, because solar lanterns are just the start of how solar can be used for powering local economies, and so development.

Vision 7: Potential

Where there are street lights, young Africans can often be seen doing what homework they can sitting beneath them. Imagine a line of children having to study below street lamps in Birmingham, England, or Birmingham, Alabama. Most African children don't even have that option, because roads are unlit and many rural homes are miles from the roads. Solar can provide electricity in homes, affordably, for hundreds of millions of households, without recourse to grids or generators.

Pilot projects in Malawi and Kenya, set up by the charity SolarAid, have shown that local groups can be trained relatively easily to convert standard, medium-sized kerosene 'hurricane' lanterns, or build free-standing solar lights. Conversion of the lanterns involves putting

A vocational training centre in Bwelero village, Malawi, with classes lit by solar energy.
(Image: SolarAid - Andy Bodycombe)

rechargeable AA batteries into the chimney and using light-emitting diodes (LEDs), which are very cheap and efficient, to direct light down onto a cone reflector, which sits over the top of the old wick. Low-cost solar lanterns sell easily, and no wonder: switching from kerosene can increase a family's net income by 20%. Market research in the field suggests people spend the extra disposable income on clothes, books, school fees, and medication and home-care for HIV-positive relatives. Subsistence farmers also have the option of spending their disposable income on seeds and fertiliser.[30]

On top of these benefits come the carbon savings. Kerosene produces 2.6 kilograms of CO_2 per litre. At an average usage of 1 litre per week, a kerosene lamp converted into a solar lantern should save 135 kilograms of CO_2 per year, or a tonne of CO_2 during the average seven-year life expectancy of the product. With solar lanterns spreading into millions of homes across the developing world, this soon sums to a useful contribution to greenhouse-gas-emissions reductions.

The next level up from the solar lantern is the solar home. In Asia, Africa and Latin America solar companies are showing that if people have access to simple microcredit, it is a relatively simple matter to deliver solar home systems to them. When the channels of credit and distribution

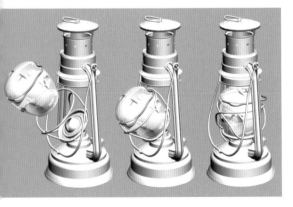

SolarAid prototype LED module for existing solar lanterns.
(Image: SolarAid

are in place, the default rate on loans for solar is very low. The Grameen Bank in Bangladesh has done much to pioneer small-scale lending to the rural poor. In recent years microcredit has been spreading, and as a consequence solar companies targeting the poor are growing. The prize is huge here: all of non-electrified sub-Saharan Africa could be provided with clean electricity from small-scale solar for less than 70% of what the wealthy industrialised countries spend on subsidies for fossil fuels every year.[31] The International Energy Agency calculates that providing basic energy requirements to every

individual in the developing world could be achieved with 16 gigawatts of photovoltaics.[32] That would cost some $80 billion, the IEA estimates: equivalent to just five months of current expenditure on the wars in Iraq and Afghanistan.[33]

Vision 7: Progress

When Bill Clinton visited the first solar-powered village in Ethiopia in 2008, he spoke of a coming revolution in solarising developing countries: 'It's the energy equivalent of the cellphone movement,' he said. The village, Rema, has 1,100 homes 'shining in the dark evenings like white beads on a string', as one British newspaper reported. The Solar Energy Foundation, set up by Harald Schutzeichel, solarised the village at a cost of £240,000, mostly supplied by Good Energies (see p.132). He has set up a solar energy school which has trained twenty-four technicians from all across Ethiopia, and is now looking for €10 million to set up a microfinance bank, from which villagers can borrow at affordable rates.[34]

When it comes to poverty and hunger alleviation, development experts increasingly tend to favour the self-help approach over the direct-aid approach. In Malawi, for example, supplying farmers with cheap seed and fertilisers has worked well. The Malawian maize harvest was disastrous in 2005, but doubled to a surplus in 2006 after targeted self-help aid. In 2007 the harvest was some 50% above the five-year average.[35]

SolarAid trainees in Malawi.
(Image: SolarAid)

Training microsolar entrepreneurs in such countries holds the potential to become a particularly effective example of the self-help approach to aid. It is easy to imagine microsolarisation using these tools quickly spreading across entire countries – or even continents, provided microfinance is readily available. In that respect, the trend is encouraging. The global microfinance movement now has more than 3,000 institutions with some $25 billion of loans in place among some 125–150 million customers. Deutsche Bank entered microfinance in 1997, Citibank in 2005, and JP Morgan in 2007. Barclays has a microfinance initiative in Ghana.[36] With these trends in place, plus the underlying 'no brainer' economics of solar lanterns, Vision 7 must surely be feasible.

Vision 8: Solar mobilising a new class of responsible investors

The financial system that most people assumed was safe, sound and a route to economic well-being proved in 2008 to be risk-laden, systemically flawed and a route to potential global economic ruin. This failure invites root-to-branch economic reform. This might once have been regarded as an extreme prospect, but today it is viewed by many as imperative – and even inevitable. That context makes this particular vision far more realistic.

The vision

Socially responsible investment of all kinds has exploded into the mainstream. Where in 2008 it was limited to minority markets (admittedly fast-growing ones) now it dominates. Equity funds are rigidly screened on the widest possible definition of sustainability, including sustainability of corporate governance and risk process. This in turn is influencing strategy in corporate boardrooms. The changes of strategic direction have included a major shift from centralised power generation to energy services and microgeneration, because this is the asset class that increasingly appeals to the new investors.

Meanwhile, innovative new forms of financing and savings have emerged. These are increasingly related to energy. Among them, solar bonds have become commonplace. In these, pension funds invest in pooled assets comprising income streams from solar roofs (electricity sold and carbon credits earned). Such bonds are making reliable returns, and they are widely acknowledged to be safer than a government bond, because the feed-in tariff payments are guaranteed, and the bond is backed by the considerable asset of the solar roof itself. The returns might not have been enough to satisfy the hedge funds of 2007, but they are quite enough for the pension funds of the 2010s.

At the citizen level, peer-to-peer lending – where people with savings seek safe returns by lending to others with income or other collateral – has become a megatrend. So too has citizen development-financing, as people with small amounts of savings in the rich countries extend microfinance loans to those in need of capitalising small entrepreneurial enterprises in the underdeveloped countries. Such flows of money are measured in billions of dollars, and the default rates have proved to be amazingly low.

Summing all these trends, as society heals itself in the wake of the global financial crisis of 2008, a kind of renaissance is emerging. Much of it has to do with the way the world uses, and finances, energy. Within that sector, the solar-energy industry is now one of the dominant players.

Vision 8: Potential

Many of the problems in energy stem from an institutional culture that has grown used to 'thinking big'. To this day, despite all the problems involving greenhouse gases and energy security, government officials often stick doggedly to the past. If there is a problem with coal power stations, they say, then let's simply try to capture their carbon emission, or build huge nuclear plants in place of huge coal plants.

The same big-thinking has applied for a long time to finance, and to development aid. The financial institutions have been quick to complain about the transaction costs of small projects and slow to apply the innovative tendencies they brought to bear on the creation of complex derivatives. Governments and aid institutions have tended to argue that developing countries need to build big projects, because the best projects in the developed world are big. And so the developing world has mostly ended up with dams and coal-fired power plants. The Kyoto Protocol's Clean Development Mechanism (CDM) illustrates this culture at work. The CDM allocates carbon fees from polluters who exceed their carbon allowances in the developed world to carbon-reduction projects in the developing world. To date, while CDM funds pour into Asia and Latin America, Africa has been allocated very little. The main reason is that it finds difficulty identifying projects big enough.

The rethink that is under way in all aspects of finance, as the global financial crisis unfolds, offers the potential to replace a fossil-fuel-centred, centralised energy model with a decentralised, low-carbon one. And therein lies the route to many prizes: climate-change abated, energy-security enhanced, and economies reflated with job-rich new industries in which solar – alone and in hybrid energy systems – figures strongly.

Vision 8: Progress

In the early 1990s, advocacy of environmental or social screening of investment was restricted to a far-sighted few, like Stephan Schmidheiny, founder of the World Business Council for Sustainable Development, and Tessa Tennant, an early leader of ethical investing in the UK. Today, following the lead of pioneers like these, a whole new class of investor has emerged, at both the ordinary and major investor levels. Small investors are increasingly turning to ethical investment products. A 2008 survey by the Co-operative Bank shows 85% of investors in ISAs (a class of equity investments in the UK) will consider using an ethical scheme, up from 67% in 2007. Almost 80% believe that such products can perform at least as well as the mainstream. At the start of 2008, the total sum invested in ethical financial products was up 15% year-on-year, standing at £13.3 billion.[37]

In March 2008, the Head of Deutsche Bank Asset Management reported that it was becoming easier to use campaigning investment as a tool in fighting global warming. Until recently, mutual funds – the most popular form of investment – offered few options. In the last two years, an estimated 200 mutual funds and exchange-traded funds have been launched targeting companies mitigating or adapting to climate change. Some $66 billion of retail-investor money has flowed into them. Wind and solar have proved particularly attractive.[38]

Famous global-warming campaigner Al Gore is now actively involved in attracting sustainable investments. He has teamed up with former head of asset management at Goldman Sachs, David Blood, to create Generation Investment. In April 2008 they announced a fund to invest $683 million to invest in early-stage environmental companies. The target will be small companies in renewables, efficiency, biofuels, biomass and carbon trading. This Climate Solutions Fund joins their Global Equity Strategy Fund: $2.2 billion for large companies in sustainable fields.[39]

Marcel Brenninkmeijer.
(Image: Good Energies)

Among the major investors, Marcel Brenninkmeijer cuts a heroic figure in terms of leadership. A member of the wealthy family owning the C & A retail chain, Brenninkmeijer saw that it was possible for solar to make money for his family while achieving considerable social benefit. He set up a fund to invest in solar and other renewables and called it Good Energies, to show where his sympathies lay. By the end of 2007, primarily by investing in solar-photovoltaic companies, he had built the fund into a value measured in billions of euros.[40] Brenninkmeijer says that his intention from the outset was to use the tools of conventional investment to do good in the world, and solar was and is extremely well positioned as a vehicle for doing that. 'If it's just about making money,' he says, 'then we don't do it.'[41] Good Energies has set up a foundation with part of its profits to promote the dissemination of solar and other 'good' energy sources in the developing world.

Rich individuals are a key driver in responsible investing, the European Social Investment Forum reports. The Eurosif survey records more than €500 billion of rich people's money in social investment in 2007 and forecasts a doubling to more than a trillion by 2012. 'Successful entrepreneurs of today are not the industrialists of yesterday,' one respondent to the survey points out. 'They are younger and more interested in sustainable investments.'[42] The decision in the English case of Cowan v Scargill in 1985 continues to be used by pension funds as authority for the conclusion that they can only invest with the best financial returns in mind, rather than taking any ethical stance.[43] However, this view has been convincingly challenged in a 2005 report to the UN by the international law firm Freshfields Bruckhaus Deringer. This report also provides authority for the contrary view that in fact failing to take ethical issues into account might constitute a breach of the pension fund trustees' fiduciary duties. This opinion has not been tested in court yet, but nonetheless funds are increasingly engaging with companies on ethical issues and screening out those with unappealing climate liabilities.[44]

The drivers for this transfer of priority lie partly in the multiplying problems of fossil fuels, partly in the growing reliability and innovation of renewables, and partly with the increasing seriousness of the global environmental crisis. The Worldwatch 'State of the World' report for 2008 observed that environmental problems, once considered irrelevant to economic activity, are now 'drastically rewriting the rules for business, investors, and consumers'. Everywhere the dynamics of business are changing, the report observes. Companies are lobbying Congress for regulation favouring renewables, and unilaterally cutting their carbon dioxide emissions. DuPont, for example, has cut its greenhouse-gas emissions by 72% from 1991 levels, saving £1.5 billion in the process. Fifty-four banks, representing 85% of global private-project finance capacity, have endorsed a new international standard of sustainability investment, the Equator Principles. The list of corporate engagement, and meaningful action, is long.[45]

Also growing fast is microfinance, a concept that first came to prominence in Bangladesh, where the Grameen Bank started making tiny unsecured loans to entrepreneurs, so that they could start small businesses. So successful was the scheme that in 2006 it won its

Muhammad Yunus, founder of the Grameen Bank.
(Image: Grameen Bank)

architect, Muhammad Yunus, a Nobel Prize. Today the idea is spreading fast. In Kenya, for example, Equity Bank offers microloans backed by unusual guarantees – groups of neighbours vouching collateral, or matrimonial beds in the case of women. It loans as little as £5 to its 3 million customers, yet has still grown to be one of the leading companies on the Nairobi Stock Exchange. Its default rate is 3%, compared to an industry average of 15%. Unlike Grameen Bank, which has used donor funding and state subsidies, Equity Bank is entirely commercial. It has more than a hundred branches, and uses armoured trucks as mobile branches in some rural areas. The typical savings account contains £100.[46]

In Silicon Valley, a group of entrepreneurs quit their mainstream jobs in 2005 to set up a non-profit organisation, Kiva.org, to facilitate interest-free lending by Americans to entrepreneurs in the south. Often the latter need tiny sums to get started. The maximum size of a Kiva loan is just $25. Ordinary Americans queue to make them: $25 means that the lender-philanthropists do not need to be wealthy themselves, and sure enough, a whole new category of donors is signing up with Kiva. The default rate at the other end is vanishingly small: just 0.14%. Many lenders don't even want their money back: they just want to give people in the developing world a leg-up. The vast majority of their returned loans go into a revolving fund to bankroll yet more entrepreneurs who want to open small shops, buy sewing machines, or whatever they plan to do to lift the poverty barrier. By March 2008, Kiva had signed up nearly a quarter of a million lenders, and loaned $22 million in forty countries. It targets a billion dollars of loans within a decade.[47] Solar projects like solar-lantern distributorships, and solar-home-system installers, are going to make particularly attractive targets for the 'new lenders'.

Whether a non-profit model works best for microfinance, or a fully commercial one, or a combination remains to be seen. But one thing is clear. This is a societal development that sits very comfortably with the solarisation that is under way. The solar century could end up having a very different set of financial paradigms from the hydrocarbon century.

Vision 9: More ambitious UN development goals – made possible by solar

At the Johannesburg summit on development and environment in 2002, governments established a number of so-called Millennium Goals for alleviating poverty. Entering the global financial crisis of 2008, they were falling badly behind the targets they had set six years before. The financial crisis itself was soon exacerbating this situation. The ninth vision, therefore, involves simply the reversal of this trend.

The vision

Widespread solarisation is greatly improving health, education, sanitation, water quality, household firewood use and gender imbalances across the developing world.

This is happening so fast that governments in the UN are able to reset their development goals, this time more ambitiously, and with every chance of achieving them.

Vision 9: Potential

Vaccinations could cut millions of deaths from diseases such as measles. But vaccines have to be kept cool – a problem in developing countries with hot climates, unreliable power supplies, expensive portable fridges and rural populations. Solar fridges can make a real difference. They can be connected to the grid – with a solar panel and battery as back-up – or can be stand-alone systems. Either way, each solar fridge can work for years and store thousands of doses at a time.

That's just one way that solar can improve health. Currently some 2.6 billion people – more than a third of the global population – lack access to proper sanitation. Unsafe drinking water and other impacts of poor sanitation kill more than a million and a half children every year from diarrhoeal diseases, and lead to countless lost school days. Renewables could help. Solar can promote sanitation via pasteurisation, desalination and the pumping of water. Pumps powered by renewable energy allow wells to be sunk closer to homes, cutting water-carrying time. Wind-powered water pumps are well known in developing countries, but increasingly solar pumps are being used. Pumps can make a huge difference to rural farmers. A small plot of maize can need up to 5,000 litres a day, and hand pumping that amount from 10 metres depth would take six hours. Groundwater is often found at depths of 30 metres. Renewable power for the pumps is the answer.

Another way to improve health significantly, especially for slum dwellers, is to combat air pollution. In urban centres, coal is burnt routinely in millions of tonnes each year, creating filthy air and emissions of greenhouse gas and acid rain pollutants. Solar could make a big difference here, just as it could help combat maternal mortality. Unicef figures show that a woman has a horrific 1 in 22 chance of dying in pregnancy or childbirth in Southern Africa, compared to 1 in 8,000 in a country like the UK. More than a third of deaths are caused by haemorrhage, and deaths from infections are also common. Proliferation of solar-powered medical centres would help enormously here, as would solar-powered water sterilisation.

Interestingly, the single biggest correlation with the birth rate is female literacy, and hence education. Simply stated, the more educated a young woman is the fewer babies she is likely to have and the less she will be at risk of death in childbirth. Educated women are also less likely to place themselves at risk of infection from HIV. The solar lanterns already discussed should go some way to helping here, and solar in general could support developing-world education more broadly. In Tanzania, for example, the majority of the newly established community secondary schools are without electricity and hence without modern training facilities. More than 2,600 community secondary schools have been built, of which more than 80% are not electrified, according to statistics from the country's education ministry.[48]

Solar could also help combat gender inequality. Currently, big gender disparities arise in Africa and Asia in large part because women tend to be tasked with the time-intensive jobs of collecting firewood, collecting water and cooking. Wood and water can often be far from villages, and women often enlist their daughters in these services from an early

Solar cookers in Peru.
(Image: www.solarcooking.org)

age. Widespread availability of modern energy would give women more time for other more productive activities, and enable more widespread education of daughters.

A third of the global population still cooks on open fires. Others use wood in stoves, often very inefficient ones. Solar cookers can be used instead, saving the long walk to and from the forest, and the chopping. Coat a big bowl in aluminium, and position a pan on a frame in the focal point of the sun's reflected rays, and you can cook pretty much anything. You can generate temperatures of up to 200° Fahrenheit, boil water in ten minutes, and cook rice in twenty minutes. Solar cookers can be found in refugee camps all over the world, and their use can be traced back to the Napoleonic era. Their indirect benefits, beyond obviating the need to cut local timber for firewood, include making it possible to boil water for pasteurisation, thereby improving health and cutting school days lost.[49]

There is also plenty of scope for solar to contribute to the development of the kind of north–south partnerships that will need to be built if global society is to get on track for the eradication of poverty through sustainable development. Much of the infrastructure to bring modern communications to the rural poor, for example, would have to be solar-powered. Many of the mobile-phone networks going into Africa have to rely on solar electricity.

Vision 9: Progress

The experience of Dr Sonexay Phonexaysack, a doctor at a solar-powered health centre in Ban Kuai village, Laos, shows the potential of the solar-powering of health centres and hospitals. 'Before we had solar,' he says, 'we had to fetch essential medicines and vaccines from elsewhere, because we had no way of keeping them cool here. Often people are very ill by the time they reach here so it could make a difference to whether they live or die. With solar, we can operate at all hours. We used kerosene lanterns before, but they were dirty and smoky and the light was poor.'[50]

A solar initiative of the Shidhulai Swanirvar Sangstha organisation in Bangladesh shows the benefits of bringing education to women in remote rural areas, in this case via floating solar-powered classrooms. Girls who were previously excluded from education because their parents couldn't afford to send them long distances or did not want them to leave the village, are now being educated on a daily basis in these classrooms. This education can include rights-based education through the Girl Children's Rights Association, a distance education programme that provides information to girls and young women on topics such as domestic abuse, child trafficking and prostitution.[51]

In the remote reaches of the Sunderbans in Bengal, where rural electrification is still a distant dream, six women are harnessing solar power to show fellow villagers how they can bring light into their lives. The women design and assemble solar-powered torches and night-lights that have become must-haves in the villages. 'We are very happy because earlier we were mere housewives,' says one. 'Now we work for two to three days a week and assemble night-bulbs. We have also learnt to market and sell these products to the people.'[52]

Developing-world solar applications aren't limited to rural areas, however. Imagine if every city, in both the developing and developed world, followed the examples of Rizhao and Kunming in China. Rizhao means City of Sunshine in Chinese, and it is easy to see why. In the central districts, 99% of households use solar water heaters. With half a million-plus square metres of solar-thermal panels in the city, costs have fallen, manufacturing plants have been set up and coal power stations have been replaced. It's no coincidence that Rizhao is among China's top ten cities for air quality. More than this, Rizhao's city leaders have come to view solar as the pivot in an increasingly positive social, economic and cultural development plan for the city.[53]

In Kunming, in the western Province of Yunnan, more than half the city's inhabitants use solar hot-water heaters.[54] The use of solar thermal in this province – along with other alternatives to traditional firewood like biogas and fuel-efficient stoves – is particularly important because mountainous Yunnan is botanically the richest temperate region in the world, and protection of its watersheds is crucial to the livelihoods of 500 million people downstream.[55] Reducing the need for firewood saves time and labour and helps combat soil erosion and desertification.

There's still a long way to go, of course. China has more solar hot-water heaters than all of the rest of the world combined – well over 40 million – but that's still only around 10% of the country's households. But these snapshots of progress in using solarisation for development give a feel of the remarkable things that are possible.

A Shidhulai boat library visits
a remote village in Raishahi,
Bangladesh. Electricity on boats
is generated by solar PV.
(Image: Ashden Awards)

Chapter 6
MAKING IT HAPPEN

How to Make the Solar Century a Reality

Almost all new technologies require their markets to be supported actively during the birth phase. The classic example of this is the microprocessor – the heart of the personal computer – which would not have quickly exploded into everyday use, at affordable prices, unless the US government had enabled mass use via its own procurement policy. Solar-photovoltaic technology had been around for several decades before governments became serious about building markets big enough to make an industry. In this chapter we review progress so far and ask what else needs to be done to make solar markets grow to critical mass.

Stimulating solar markets

The process of serious commercialisation of solar photovoltaics began in the 1990s, primarily in Japan and Germany. Japan elected for the subsidy route, awarding 50% capital grants for the up-front cost of solar systems, decreasing in small percentages each year as the price fell. The government reasoned, rightly, that the price would surely fall because the increased volumes would allow larger manufacturing plants, capturing economies of scale. Their policy allowed Sharp to become the largest solar-cell maker in the world, with other Japanese companies not far behind. In no time at all these companies were exporting as well as servicing the fast-growing domestic market.

The Germans started a little later and chose the feed-in tariff route. Feed-in tariffs work by charging a small surplus on everybody's electricity bill, the surplus is used to create a pot of money from which premium prices are paid for solar electricity fed into the grid. This has the same effect as a subsidy, making the solar system an attractive investment even for people unconcerned about the environmental benefits.

At least sixty countries have policies actively promoting renewables, and at least thirty-seven of these have opted for feed-in tariffs, mostly

Opposite:
Solar architecture in the UK –
the Eden Project in Cornwall.

since 2002. Some countries, such as France and Italy, give more generous feed-in payments for building-integrated solar than they do for ground-mounted solar, in an effort to steer the market in the direction of the built environment. Renewable portfolio standards are also common, both at national and state levels. These are obligations imposed on power providers to have a particular mix of renewables. Twenty-five American states have also chosen this route.

Other market-stimulation policies include direct capital investment subsidies or rebates, tax incentives and credits, sales and other tax exemptions, and net metering. At least thirty-five countries – and half of the American states – offer direct capital investments, grants or rebates.

Spanish solar PV in Seville.
(Image: www.abengoa.com)

Perhaps the most impressive is California's Solar Initiative, which is aiming for 3 gigawatts of solar on buildings by 2017, and will deploy $3.1 billion in subsidies along the way.

To stimulate solar-thermal markets, governments are increasingly opting for mandates. These stipulate a minimum amount of solar water-heating in new properties. Israel was the first to mandate solar thermal in all new builds, in 1980. Spain followed in 2006 with a mandate for both thermal and PV in certain types of new construction and renovation. At least four other countries introduced mandates in 2007: India, China, Korea and Germany.[1]

Local government has been very active in stimulating solar thermal and other renewables markets. Over seventy municipalities in Spain introduced solar-thermal ordinances prior to the national mandate. Nagpur in India has a mandate. Rome is the latest major city to join the club.[2] Many cities have declared carbon dioxide-reductions goals ahead of national action, and some of these require renewables in the energy mix. Over 100 British municipalities require 10% energy generation or carbon reduction onsite in new buildings from renewables.

A start has been made on the solar road, as a result of the measures described. Momentum has been created. To keep it up, it is vital that leaders in government set ambitious but achievable targets, and this too is beginning to happen.

Setting targets

By 2008 there were renewables targets in at least sixty-six countries around the world. Most of them are for shares of national electricity production, typically in the 5–30% range. Much more ambitious targets, that might once have seemed unrealistic, are today beginning to seem achievable. In this respect, Al Gore has called for a national US mission to generate 100% of the country's electricity from non-fossil sources within ten years. He has appealed to President Obama to emulate John F. Kennedy's Apollo mission. 'We are borrowing money from China to buy oil from the Persian Gulf to burn it in ways that destroy the planet,' Gore says. 'Every bit of that has got to change.' Gore insists the 100% target is 'achievable, affordable and transformative'. He argues that: 'When we send money to foreign countries to buy nearly 70 per cent of the oil we use every day, they build new skyscrapers and we lose jobs. When we spend that money building solar arrays and windmills, we build competitive industries and gain jobs here at home… This is a generational moment.'[3]

At the time of writing, the Obama administration is beginning to execute the biggest economic stimulus programme in US history, aiming to create 3 million jobs, with wind, solar and a smart grid high on the list. The $675 million pledged is far higher than the programmes in the UK, China and Japan, and it is much needed. 1.5 million American jobs were lost in 2008, with 4 million more at risk, threatening a total of 9% unemployment. The world is watching, and hoping.

American states are endeavouring to do their bit. Governor Schwarzenegger has signed an executive order with a target for California of 33% renewable energy by 2020: America's most aggressive state target. Schwarzenegger is drafting legislative language to require all utilities to meet the target and spread the cost across all ratepayers with safeguards for people on low incomes. The state also plans to create a one-stop permitting process to halve application time for renewable projects.

In Europe, twenty-seven EU leaders agreed to a '20:20:20' climate deal in December 2008: 20% emissions cuts, 20% renewables in the energy mix, and 20% improvement in energy efficiency, all by 2020. If the energy target was met purely by building renewable electricity capacity, that would imply a renewables percentage in electricity of 34%. Currently the share of renewables in the energy mix of European countries ranges from almost nothing in Malta to 40% in Sweden. As for the other tool of which the European governments have such hopes, the EU's Emissions Trading Scheme, huge loopholes have been created by exemptions for heavy industry. The EU can no longer claim the scheme will bring major emissions reductions, or that the US and the rest of the world will follow. This throws even more emphasis onto renewables, in terms of the heavy-lifting work needed in cutting emissions.

As in the US, ambitious target-setting is resonating at the state level too. The central Spanish state of Castile-La Mancha, mentioned in this book's introduction, has a population of 2 million and a passion for renewable energy. In 2006 it drew up a plan to provide 100% of its energy from renewables by 2012. Entering 2008, it had reached 40%. Unsurprisingly, Castile-La Mancha is the number one Spanish state in terms of installing photovoltaics.[4]

Overcoming failure of imagination: a new constituency emerges

In the solar industry, trade shows that looked like cottage-industry fairs a few years ago now host giants of the digital revolution selling entire solar factories that will soon have gigawatt-a-year production capacities. Investment has been flooding into the renewables sector, not least from Silicon Valley. A sixth of the world's electricity and a third of new electricity now come from micropower,[5] rather than from central fossil-fuel stations. In a dozen industrial countries, micropower now provides between a sixth and over a half of all electricity. Micropower added up to 58 gigawatts in 2006. In terms of private-risk capital investment, $56 billion went to distributed renewables.[6] Electricity utilities increasingly tend to go through the motions on nuclear and coal, meanwhile placing ever bigger bets on renewables. Similarly, car companies are showing a clear preference for electric battery vehicles and hybrids, and consultants are competing with each other to design the infrastructure for smart grids.

With this tide of change, a new constituency is emerging, and growing. In 2000, most true believers in a renewable-powered future were environmental campaigners. Today, true believers span a wide spectrum, including right across the business world. For example, in the UK, the Co-operative group has taken out full-page advertisements in national newspapers announcing: 'We've put £1 million into solar power for schools. How's that for a bright idea?' The smaller print talks about how 'we aim to be good for everyone' – the mission message. 'We hope that, as well as producing clean electricity, the Green Energy for Schools scheme will produce more enlightened children.' There is no overt appeal in these advertisements to bank, or take out insurance policies, with the institution responsible for the ads.

In the US, Bank of America's CEO has called for 'a new economic

Pupils at the 100th school in the UK to be fitted with a solar PV system by the Co-operative group.

Solar PV under the familar
Tesco logo, in Swansea
& N. Ireland.

future based on clean, renewable energy'. The markets will not get us there, Ken Lewis says. Governments must intervene, he concludes, no matter how strange it seems for a bank CEO to say that. 'It has become clear that there is a strong connection between our willingness to diversify our energy sources and our ability to grow the economy sustainably.' He calls for tax credits to be renewed. With these kinds of incentives, banks can do more leasing of solar panels, for example, relieving consumers of the need for up-front capital purchases. Effectively, he is calling for a green new deal, though he doesn't use those words.[7]

At the same time, supermarkets are competing for title of greenest retailer. Asda boasts about a new 'low-carbon store' in Liverpool that has doors on the refrigerators, one obvious (and long resisted) measure saving 8% of emissions in a store that will cut 50% of emissions overall. The title seems to change hands regularly. Sainsbury's opened a store in Dartmouth that cut emissions by 40%. Then Tesco trumped that with a store in Shrewsbury that cut emissions by 60%. With 40% of a store's costs being electricity, energy bills are now a driver alongside climate-change performance. All UK retail chains now have CO_2 targets that go beyond tokenism, with Tesco aiming to halve emissions from existing stores and distribution centres worldwide by 2020. They feature solar strongly in their social-responsibility narrative.[8]

Another giant, P&G, gives us a clue as to how far-reaching the response to the end of cheap oil will be in the business world. This massive manufacturer, with 145 plants in eighty countries supplying 3.5 billion consumers, plans to shift to factories close to customers in order to cut its fuel bills. The company is also investing in renewables. By the end of 2009, half the electricity at a Pennsylvania nappy plant will come from onsite wind, for example, and other schemes are being rolled out, including solar projects.

Even electricity utilities are active in the solar-for-change game.

In the UK, npower took out a front page advert in a national newspaper promoting solar. 'npower solar works even when our boys can't,' the slogan reads. The advert, showing a picture of a cricketer with bat raised, is a play on the dismal performances of England's cricket team. Cynics may dismiss this as tokenism, but in a world where some of their biggest clients – banks and retailers – are so clearly going well beyond tokenism, it would be a foolish utility that assumed the past was going to equate in any way to the future where solarisation is concerned.

In case there was any danger that they would, the main solar photovoltaics trade magazine organised a conference in California in October 2008 to point out the danger to them. The organisers argued that solar PV markets are poised to grow so fast that they pose the danger of a 'black swan' event for utilities: an event that they consider highly improbable that is actually not. 'Our frank assessment is that rapidly expanding PV has significant potential to disrupt traditional electricity companies,' the organisers wrote in their report. By 2012, they argue, annual additions of solar PV could amount to 137 gigawatts, fully a third of all annual global electricity capacity additions, so creating economic, operational and financial challenges for traditional electricity.[9] Of course, such a change would also offer opportunities for utilities with foresight, and the ability to take positive proactive action.

A former CIA director, Jim Woolsey, has an explanation for what is going on. Two great drivers are gathering behind renewables markets, he argues. One is global warming. The other is energy security. You can believe in one, or the other, or both. You would be pretty foolish if you believed in neither. Jim is not alone in this view. As the CEO of Renault-Nissan put it recently, when announcing that his company would be focusing on electric battery vehicles: 'We must have zero-emission vehicles. Nothing else will prevent the world from exploding.'

In the media, the *The Economist* is among the many organs to understand what is going on now. Not long ago they scoffed at it all. Writing about the Berlin climate summit in 1995, an editorial advised: 'Most actions (to cut carbon emissions) would pose a bigger threat to human well-being than does global warming.' In June 2008, by contrast, we read the following about the prospect of replacing fossil fuels with clean energy: 'Such a failure of imagination (that we can't do it) has been at the heart of the debate about climate change.' An *Economist* special report on the future of energy is written throughout with optimism of the kind we espouse in this book. 'Some think alternative energy will be the basis of a boom bigger than information technology,' the editorial concludes.

Running economies: the new priorities

The argument about how much it will cost to transform the powering of the global economy from carbon-based energy echoes on today in many fora. In the climate negotiations, President Bush's negotiators long argued that even small cuts in emissions would cost so much that they would imperil the US economy. Many in the US and elsewhere argue precisely the reverse: that small cuts save money, and that deeper cuts can be afforded. American energy-efficiency guru Amory Lovins and his many followers point out that in the modern world it is far cheaper to save or substitute a barrel of oil than it is to produce it – and that continuing to stoke climate change and remain dependent on imported oil and gas will torpedo our economies and cancel out any growth fuelled by fossil fuels.[10]

But even accepting the arguments of the Lovins camp, it is certain that budgetary constraints will tighten as the financial crisis plays out. Therefore new areas of budget savings would be very useful in providing the capital for energy conversion. Where could they come from?

One of the most obvious budgetary savings from a switch to solar and related clean-tech is in the area of the military. When we cease to covet oil then many of the current tensions in the world will ease and in that way solarisation actually has a part to play in reducing military spending and aiding stability. Nobel Prize-winning American economist Joseph Stiglitz has calculated the cost of the Iraq and Afghanistan wars at $16 billion a month, a sum equivalent to the entire annual budget of the UN. This is on top of regular Department of Defense expenses. In 2008 Stiglitz published a book entitled *The Three Trillion Dollar War*, referring to the US costs alone to date. The UK and the rest of the world shoulder around a further $3 trillion in costs. The current operating costs of the war cost each US household $138 every month. The annual cost to the US of the rising price of oil was running at $25 billion early in 2008, and a projected extra $1.6 trillion over the period to 2015. Much cash goes missing, including the infamous $8.8 billion for the Development Fund for Iraq. By comparison, America is currently spending $5 billion a year in Africa. Americans will

have paid $1 trillion in interest alone on the money borrowed to fight the war by 2017: a couple of hundred billion a year. As President Bush cut taxes at home while waging costly war, so he had to borrow more, often from China.

For people worried by climate change and the imminence of a global energy crisis, these kinds of budgets are incomprehensible. It is easy to imagine how even a fraction of these sums – even if the calculations are on the high side – could accelerate the process of decarbonising the US economy.[11]

But the US is not alone in facing massive national budgets directed in entirely the wrong direction when it comes to survival. Total global military expenditure in 2006 was $1.2 trillion, around 2.5% of the world's $48.5 trillion global GNP.[12] Arms-control agreements and other improvements of common security that free up military expenditures for environmental and/or energy security are arguably going to avoid far more economic damage than military action can, no matter how well thought through.

Where might other non-military budget savings come from? The UK budget in 2008, billed as the greenest budget ever, barely tinkered at the margin with climate change. The finance consultancy KPMG observed

Green collar workers en route to Number Ten to petition the UK Prime Minister, April 2009

that the measures, centring on taxing polluting vehicles, might reduce greenhouse-gas emissions 5% by 2015 – way off track – while raising less than £2 billion per year. The proportion of revenues being raised by green taxes actually fell marginally in the UK in 2008–9. KPMG's head of environmental taxes and incentives said: 'It is still very unclear how the vast majority of carbon reduction will be achieved.'[13] This is true of many nations.

Environmental taxes as a proportion of total receipts are around 10% in the Netherlands and Denmark, and less than 7% in the UK.[14] This has to change. We must put a price on carbon and other forms of pollution if we are to head off the worst effects of climate change and the depletion of oil and gas reserves. If we do this, and at the same time demilitarise budgets to the maximum extent possible, we will free up plenty of cash with which to modernise our economies, and manage our budgets.

There would also be considerable savings. These start with energy-import bills avoided, so strengthening balance-of-payments positions in importing nations. There would also be considerable savings in healthcare. Burning fossil fuels puts dangerous pollutants into the air, including nitrogen oxides, sulphur oxides, hydrocarbons, dust, soot and other suspended matter. These pollutants can cause serious health problems. According to the World Health Organisation, 2.4 million people die each year from causes directly attributable to air pollution. Worldwide, more deaths per year are linked to air pollution than to car accidents, with research published in 2005 suggesting that 310,000 Europeans die annually from air pollution. Direct causes of deaths related to air pollution include aggravated asthma, bronchitis, emphysema, lung and heart diseases, and respiratory allergies. According to the American Lung Association, switching to electric vehicles in California alone would save the US more than $2 billion each year in health bills from exposure to particulates. Hundreds of premature deaths and cases of chronic bronchitis could be avoided, as well as thousands of asthma attacks.[15]

Enforced air-quality standards, like the Clean Air Act in the United States, have reduced the presence of some pollutants. But historically the control of and general attitude towards pollution has not been that of prevention and innovation, but reduction where possible. This isn't good enough.

The good news is that solar power is ready and waiting to generate clean onsite electricity both in the developed and the developing worlds. Solar also has a major role to play in the charging of electric vehicles. The process of solarisation is going to save many millions of lives just from avoided air pollution, even before we begin counting the lives saved as a result of the negated impacts of climate change and conflict related to energy resources. On top of this will come financial savings, direct and indirect, from healthcare costs saved and healthy working days created.

Redesigning capitalism: the race begins

At the annual World Economic Forum of 2009, Western business and political leaders were thoroughly humbled. In 2007, the air had been full of optimism about the global economy. In 2008, business leaders worried about inflation. In 2009, however, gloom was pervasive. Chinese Premier Wen Jiabao and Russian Prime Minister Vladimir Putin mocked Western

The UK Environment agency HQ, with solar PV louvres.

leaders, Wen railing against the 'blind pursuit of profit', and Putin reminding the G7 that just twelve months before they had talked of the US economy's 'fundamental stability and cloudless prospects'. Meanwhile, a survey in the wake of the financial crash found that trust in business had plunged to just 38% in the US, down from 58% the previous year. It also showed that only 49% of Americans now think the free market should be allowed to function independently. [16]

Circumstances and statistics like these suggest that publics will expect a far-reaching inquest into what went wrong during the run-up to the crash. Politicians too have expectations in this regard. French President Nicolas Sarkozy has initiated a project that will see twenty-four eminent economists, led by Nobel laureates Amartya Sen and Joseph Stiglitz, to report on ways of improving the keeping of national accounts. The aim is to take in vital indicators currently excluded from the balance sheet, not least environmental degradation and quality of human life.

The standard economic indicator in use today is gross domestic product, or GDP. This is a measure of national income calculated by totalling all the goods and services produced in a country in a year. By this measure, government spending on prisons counts the same as spending on schools. Cleaning up an oil spill counts the same as installing solar power. Extracting oil counts as addition to national wealth, not depletion of assets. The best thing a person can do for their country's GDP is to be seriously ill on expensive medication while undergoing an expensive lawsuit of some kind, meanwhile driving around in an SUV, burning up as much oil as possible.[17] Unsurprisingly, growing numbers of people consider gross domestic product to be a completely dysfunctional way of measuring the well-being of a society. And yet even in the financial crisis, every index of progress is measured in terms of efforts to grow GDP.

Whatever its shortfalls as a measure of holistic well-being, GDP does give some indication of where wealth and resource-use are concentrated. Figure 23 also shows how inequitable that situation is. Globally, GDP is stuffed into three centres: the US, Europe and Japan/South Korea. The rest of the world lags far behind. Yet it is in the rest of the world that the majority of human beings live. If the world is to be just as well as ecologically sustainable, poverty alleviation clearly has to be tackled in parallel with the other global crises. The opportunities that the financial crash presents for redesign of the global system lends hope to people

Opposite: Solar facade system
in Derby, UK.

who believe that – for global equality and the global environment alike – history does not have to be destiny.

The 'new economics' commission set up by President Sarkozy is just one example of an increasing awareness among economists and policymakers that the pursuit of economic growth at all expense is not the wisest path for humanity to take. This new paradigm holds considerable scope for acceleration of the solar revolution, which by its very nature embodies ideals such as sustainability and well-being.

But meanwhile the bad news continues to clock up. The World Bank calculates that thirty-nine countries have lost 5% or more of their wealth because of factors related to climate change. These include depletion of non-renewable resources, unsustainable forest harvesting, and damage from carbon emissions. For ten countries, the damage ranged as high as from 25 to 60%. There is growing evidence, in other words, that the global economy is destroying its own ecological base.[18]

With the economic downturn has come concern that green business will suffer. Global sales of environmental industries are over $500 billion today: as big as aerospace or pharmaceuticals, according to the Environmental Industries Commission, which expects sales of $700 billion by 2010. But how badly will the financial crisis knock these industries back? At the time of writing, this is not clear.[19]

Global human society is involved in a race against time. The rate of capital flows into clean-energy markets in the years ahead will have a lot to do with whether we can win the race. So too will the extent to which we can engineer a green new deal, rethinking capitalism as necessary along the way. It is good news that the race is under way, but the diversion of capital from fossil fuels to solarisation has to accelerate if the vision put forward in this book is to become reality, and if this century really is to become the solar century.

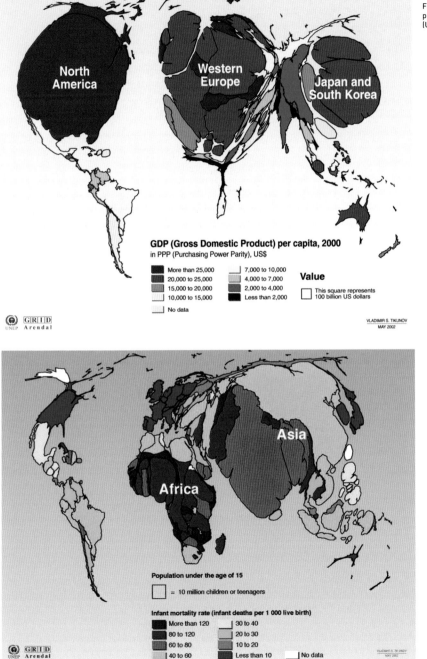

Figure 23: GDP versus population distribution. (UNEP/Grid Arendal)

Endnotes
& References

Introduction

1. Introduction 'Home, Green Home', *The Economist*, 6 November 2008.

2. "Global Trends in Sustainable Energy Investment 2008," United Nations Environment Programme: "Spending on renewables accelerates," Fiona Harvey, *Financial Times*, 2 July 2008.

3. "Cash cows on diet," Jeremy Herron, *Photon magazine*, January 2009.

Chapter 1

1. The basic unit of energy is a joule (J). To understand this, we need to begin with the force needed to change the motion of an object, measured in newtons. One newton is the force needed to accelerate a mass of 1 kilogram at a rate of 1 metre per second. Expressing this as an equation, Force (N) = Mass (kg) x Acceleration (ms^{-2}). One joule of energy is the amount supplied by a force of 1 Newton in causing movement through a distance of 1 metre. Expressing this as an equation, Energy (J) = Force (N) x Distance (m). A watt is a power of one joule per second.

2. For further information on the subjects in Chapter 1: James F. Luhr (Editor-in-Chief), *Earth: the Definitive Visual Guide*, Dorling Kindersley, 2004.
Claudio Vita-Finzi, *The Sun: A User's Manual*, Springer, 2008.

Chapter 2

1. Eric Martinot, 'Renewables 2007 Global Status Report', Renewable Energy Policy Network for the 21st Century, 2008.

2. Energy demand figures are for 2006, from 'World Energy Outlook 2008', International Energy Agency, Paris. Reserves figures are from 'BP Statistical Review of World Energy 2008', BP, June 2008.

3. 'Intergovernmental Panel on Climate Change AR4 Report', 2007, Table 4.2. They quote figures in exajoules, a calorific value, which we convert to million tonnes of oil-equivalent at 1 mtoe = 42 gigajoules = 0.042 exajoules.

4. 'The Price of Steak', National Geographic, June 2004, p. 98. The article cites a 1,250 lb steer requiring 283 gallons. 1 barrel = 42 US gallons, 6 barrels = 252 gallons.

5. Fully referenced detail on all points made about the ITPOES report can be read in the report itself, 'The Oil Crunch: Securing the UK's Energy Future', available as a pdf from www.peakoiltaskforce.net

6. For more detail see: Jeremy Leggett, *'Half Gone: Oil, Gas, Hot Air and the Global Energy Crisis'*, Portobello, 2005.

7. Winnie Zhu, 'China Halts Coal-to-fuel Projects to Conserve Coal Supplies', Bloomberg, 29 August 2008.

8. Patrick Wintour, 'Minister: We Must Build Kingsnorth To Get Clean Coal', *Guardian*, 9 August 2008.

Chapter 3

1. Shell International, 'Energy Needs, Choices, and Possibilities: Scenarios to 2050'

Shell Special Publication, 2001.

2. 'Using Energy More Efficiently: an Interview with Rocky Mountain Institute's Amory Lovins', The McKinsey Quarterly, 2 July 2008.

3. Dominik Sollman, 'Nocturnal Solar Power', *Photon Magazine*, November 2008.

4. See, for example, Amory Lovins, E. Kyle Datta, Odd-Even Bustnes, Jonathan G. Koomey and Nathan J. Glasgow, *Winning the Oil Endgame: Innovation for Profit, Jobs, and Security*, Earthscan, 2004.

5. David Strachan, 'Hydrogen's Long Road to Nowhere', *New Scientist*, 29 November 2008.

6. Kevin Allison, 'Venture Capitalists Boost Smart-Power Grid Move', *Financial Times*, 21 August 2008.

7. David R. Mills and Robert Morgan, 'A Solar-powered Economy: How Solar Thermal Can Replace Coal, Gas and Oil', *Renewable Energy World*, May–June 2008, pp. 59–69.

8. Ken Zweibel, James Mason and Vasilis Fthenakis, 'By 2050 Solar Power Could End US Dependence on Foreign Oil and Slash Greenhouse Gas Emissions', *Scientific American*, January 2008, pp. 48–57.

9. http://biopact.com/2007/12/germany-is-doing-it-reliable.html
Background paper: 'The Combined Power Plant', available on http://www.kombikraftwerk. de (F), and Christoph Podewils, 'A Reliable Ten Thousandth', *Photon Magazine*, December 2007.

10. John Vidal, 'World's Biggest Solar Farm at Centre of Portugal's Ambitious Energy Plan', *Guardian*, 6 June 2008.

11. 'The Power and the Glory: A Special Report on Energy', The *Economist*, 21 June 2008.

12. 'World publics say oil needs to be replaced as energy source', 17 April (posted). http://www.worldpublicopinion.org/pipa/articles/btenvironmentra/474.php

13. John Thornhill, 'Western Fears on Russian Energy', Paris, 18 February 2008.

Chapter 4

1. For a more detailed lay description of how a cell works see Deutsche Gesellschaft für Sonnenenergie, *Planning and Installing Photovoltaic Systems: a Guide for Installers, Architects and Engineers*, Earthscan, 2008. For a more advanced scientific explanation, see John Twidell and Tony Weir, Renewable Energy Resources, 1986.

2. Charles Yonts, Simon Powell and Ming-Kai Cheng, 'Solar Maximum: a Sustainable Source of Investment Returns', CLSA Asia-Pacific Markets, Special Report, May 2007.

3. 'Solar Annual 2008: Four Peaks', Photon Consulting, 2008.

4. Hilary Flynn, 'No Turning Back Now', *Photon Magazine*, December 2008.

5. Michael Rogol, Photon Consulting, personal communication, 28 February 2009.

6. Ibid.

7. Ibid.

8. Charles Yonts, Simon Powell and Ming-Kai Cheng, 'Solar Maximum: a Sustainable Source of Investment Returns', CLSA Asia-Pacific Markets, Special Report, May 2007.

9. Barclays Capital Daily Solar News, 21 January 2009.

10. Personal communication from CEO Mike Ahearn, 27 February 2009.

11. See the summary of a thin-film researchers' gathering in 'Optimism Despite Questions', Sun & Wind Energy, volume 1, January 2009; also Olga Papathanasiou, 'The first 20 percent Thin-film Cell', *Photon Magazine*, October 2008.

12. 'Solar Annual 2008: Four Peaks', op. cit.

13. This section compiled with information form Citigroup, Smith Barney, Solar Catalyst Group, and REC.

14. See the summary of a thin-film researchers' gathering, op. cit.

15. Anne Kreutzmann and Michael Schmela, 'Emancipation from Subsidy Programmes', Photon Magazine, December 2008.

16. 'Progress, slice by slice', *Sun & Wind Energy*, volume 6, 2008.

17. Vishal Shah, 'Solar Energy: Industry Overview', Barclays Capital Equity Research, 1 December 2008.

18. Ines Rutschmann, 'A Country of Megawatt Parks', *Photon Magazine*, September 2008.

19. David Biello, 'Solar Utility: Electricity from Sunshine on a Massive Scale in California', *Scientific American*, 15 August 2008.

20. Garret Hering, 'Flying High', *Photon Magazine*, September 2008.

21. Jochen Siemer and Garret Hering, 'Gearing Up for Price War', *Photon Magazine*, October 2008.

22. http://www.energy.gov/news/4503.htm

23. Garrett Herring, 'Dawn of 500 Suns', *Photon Magazine*, November 2008.

24. 'CPV systems will be able to compete with traditional PV systems by late 2009', *Sun & Wind Energy*, volume 5, 2008.

25. 'Solar Heating on a Grand Scale', *Sun & Wind Energy*, volume 5, 2008.

26. 'World Energy Outlook 2008', International Energy Agency, Paris, 2008, cites 6 GW in 2006 (p.168). Photon's 2008 Solar Annual gives 2007 production as 3.9 GW and 2008 production as 7.1 GW, most of which will have been installed, making 15 GW a conservative estimate for 2008 capacity.

27. 'World Energy Outlook 2008', op. cit., cites 354 MW in 2006, and 2 GW visible by 2010 (p.170). 1 GW seems the best estimate for 2008.

28. 'China's Thermal Market is the Largest', Sun & Wind Energy, volume 5, 2008. The 2006 figure was almost 128 GW, increasing at 17% p.a. on 2005. Extrapolating would give a 2008 figure of c.170.

29. 'Green Jobs: Towards Decent Work in a Sustainable Low-carbon World', United Nations Environmental Programme, September 2008.

30. www.power-technology.com

31. Statistics in this section are from: Christoph Podewils, 'Keep Your Friends Close', *Photon Magazine*, November 2008.

32. 'PV FAQs: What is the Energy Payback for PV?', US Department of Energy, January 2004.

33. Volker Buddensiek, 'Modules with a minimal CO2 footprint', *Sun and Wind Energy*, March 2009.

34. Cyril Sweett and Faber Maunsell, 'Definition of Zero-carbon Homes and Non-domestic Buildings', Department of Communities and Local Government, Consultation, December 2008.

35. 'Energy Requirements Of Desalination Processes', Encyclopedia of Desalination and Water Resources (DESWARE), UNESCO Encyclopedia of Life Support Systems (EOLSS).

Chapter 5

1. Information in this section is from 'The Elusive Negawatt: Energy Efficiency Briefing', The Economist, 10 May 2008; and 'Curbing Global Energy Demand Growth: the Energy Productivity Opportunity, McKinsey Global Institute, May 2007.

2. Martin Wainwright, 'The street in Leeds that leads the way to greener living', 29 July 2008.

3 'Green Jobs: Towards Decent Work in a Sustainable Low-carbon World', United Nations Environmental Programme, September 2008.

4. Ibid.

5. 'Economic Stimulus: the Case for "Green" Infrastructure, Energy Security and "Green"

Jobs', white paper by DB Advisors, Deutsche Bank Group, November 2008.

6. SAP's 'regulated' electricity figures only include consumption from standard appliances such as lighting and a washing machine.

7. "2020: a vision for UK PV," UK PV Manufacturers Association, Special Report, April 2009

8. 'The Growth Potential for Onsite Renewable Electricity Generation in the Non-domestic Sector in England, Scotland and Wales', Element Energy Report published by the Department of Business and Regulatory Reform as part of the RES consultation, 15 September 2008.

9. 'Solar Houses: On the Way to Mass Market', *Sun & Wind Energy*, volume 3, 2007.

10. 'China's Solar-thermal Market is the Largest', *Sun & Wind Energy*, volume 5, 2008.

11. Stephanie Rosenbloom, 'Giant retailers look to sun for energy savings', *New York Times*, 10 August 2008.

12. David Biello, 'Solar Utility: Electricity from Sunshine on a Massive Scale in California', *Scientific American*, 15 August 2008.

13. Alok Jha, 'Solar power from Saharan sun could provide Europe's electricity, says EU', *Guardian*, 23 July 2008.

14. Alok Jha, 'Power in the desert: solar towers will harness sunshine of southern Spain', *Guardian*, 24 November 2008.

15. 'Andasol 1: Heat Storage in Operation', *Sun & Wind Energy*, volume 1, 2009.

16. Todd Woody, 'Land grab', *Fortune*, 21 July 2008.

17. McKinsey & Company, 'The Economics of Solar', June 2008

18. Lazard Freres, 'Levelized Cost of Energy Analysis', February 2008. The estimates for PV reflect the current production tax credit, and assume accelerated asset depreciations are applicable. They use 2008 dollars, 60% debt at 7% interest rate, 40% equity at 12% cost, a 20 year economic life, 40% tax rate, and 5–20 year tax life. The coal estimates assume a £2.50 price per MMBtn and the natural gas a price of $8 per MMBtn.

Notes on table: Reflects production tax credit, investment tax credit, and accelerated asset depreciation as applicable.
Assumes 2008 dollars, 60% debt at 7% interest rate, 40% equity at 12% cost, 20-year economic life, 40% tax rate, and 5-20 year tax life.
Assumes coal price of $2.50 per MMBtu and natural gas price of $8.00 per MMBtu.
(a) Low end represents single-axis tracking crystalline. High end represents fixed installation.
(b) Represents a leading solar crystalline company's targeted implied levelised cost of energy in 2010, assuming a total system cost of $5.00 per watt. Company guidance for 2012 total system cost of $4.00 per watt would imply a levelised cost of energy of $90 per MWh.
(c) Represents the leading thin-film company's targeted implied levelised cost of energy in 2010, assuming a total system cost of $2.75 per watt. Company guidance for 2012 total system cost of $2.00 per watt would imply a levelised cost of energy of $62 per MWh.
(d) Low end represents solar tower. High end represents solar trough.
(e) Represents retrofit cost of coal plant.
(f) Estimates per National Action Plan for Energy Efficiency; actual cost for various initiatives varies widely.
(g) High end incorporates 90% carbon capture and compression.
(h) Does not reflect potential economic impact of federal loan guarantees or other subsidies.
(i) Based on advanced supercritical pulverized coal. High end incorporates 90% carbon capture and compression.

19. John Reed and Fiona Harvey, 'An Industry charged up: electric vehicles are poised to go mainstream', *Financial Times*, 27 May 2008.

20. 'Charge!', *The Economist*, 10 May 2008.

21. Rosie Boycott, 'We test drive the green rocket – the revolutionary sports car powered by

mobile phone batteries', *Daily Mail,* 16 May 2008.

22. www.tesla.com

23. Jonathan Soble, 'Electric cars power ahead in Japan', *Financial Times,* 26 August 2008.

24. John Reed and Fiona Harvey, 'Israel relies on electric cars to cut oil imports', *Financial Times,* 21 January 2008.

25. Alok Jha, 'Solar plant yields water and crops from the desert', *Guardian,* 3 September 2008.

26. Simeon Kerr, 'Work starts on $22 bn carbon-neutral city', *Financial Times,* 11 February 2008.

27. Karl-Erik Stromstra, 'Zero-carbon City Solar Park to Connect to Grid', *Recharge,* 23 January 2009.

28. John Vidal, 'Reaching new heights', *Guardian,* 30 January 2008.

29. John Vidal, 'Banks meet over £40bn plan to harness power of Congo river and double Africa's electricity', *Guardian,* 21 April 2008.

30. The market research data in this paragraph and the two before comes from SolarAid's experience in the field in Malawi.

31. 'Approaching global crisis threatens to reverse global development', *Environment Times,* 22 June 2004.

32. www.iea-pvps.org/pv/sa_syst.htm

33. Using the estimate of Nobel Prize-winning economist Joseph Stiglitz of $16 billion per month: Aida Edemariam, 'The true cost of war', *Guardian,* 18 February 2008.

34. Sarah Boseley, 'Power to the people', *Guardian,* 11 August 2008.

35. Harvey Morris, 'Economist plants seeds for Africa's "Green Revolution"', *Financial Times,* 19 April 2008.

36. Tim Harford, 'Are loans at 100 percent APR good for the poor?' *Financial Times,* 6 December 2008.

37. Mark Milner, 'Increasing numbers of investors turn to ethical products', *Guardian,* 11 February 2008.

38. Kevin Parker, 'Investment is key in climate change battle', *Financial Times,* 24 March 2008.

39. Fiona Harvey, 'Gore fund roots for green investing "resilience"', *Financial Times,* 30 April 2008.

40. 'We Are Not Gamblers: Interview with Marcel Brenninkmeijer and Sven Hansen of Good Energies', *Photon Magazine,* March 2008, pp. 68–69.

41. Marian Mazdra, 'Eco-Catholic as Shareholder', *Photon Magazine,* March 2008, pp. 68–69.

42. Sophia Grene, 'New era for sustainable investing', *Financial Times,* 1 September 2008.

43. Cowan v Scargill [1984] 2 All ER 750

44. "A legal framework for the integration of environmental, social and governance issues into institutional investment" October 2005. A report to UN Environment Programme (UNEP), Finance Initiative's Asset Management Working Group by Freshfields Brukhaus Deringer.

45. Paul Eccleston, 'Global warming "changing world economy"', *Daily Telegraph,* 8 January 2008.

46. Xan Rice, 'Three million customers and still counting: the bank getting rich by helping the poor', *Guardian,* 2 January 2009.

47. Jeffrey O'Brien, 'The Only Nonprofit That Matters', *Fortune,* 3 March 2008.

48. From SolarAid in-country research.

49. Anna Metcalfe, 'Sunny side up', *Financial Times Magazine*, 2 February 2008.

50. Sunlabob Renewable Energies Ltd. Laos htttp://www.ashdenawards.org/winners/sunlabob

51. Shidhulai Swanirvar Sangstha, Bangladesh Solar powered boats bringing education and sustainable energy to remote areas: http://www.shidhulai.org/ and http://www.ashdenawards.org/winners/shidhulai

52. http://www.ndtv.com/features/showfeatures.asp?slug=Solar+power%3A+Bengali+women+usher+change&Id=1212

53. Xuemei Bai, 'China's Solar-powered City', in State of the World 2007: Our Urban Future, Worldwatch Institute, May 2007.

54. Ryan Hodum, 'Kunming heats up as China's "solar city"'. www.worldwatch.org/node/5105, 5 June 2007.

55. www.c-reed.org/EN/background/index.htm

Chapter 6

1. The information in this section is from Eric Martinot, 'Renewables 2007 Global Status Report', Renewable Energy Policy Network for the 21st Century, 2008.

2. Ina Ropke, 'Solar Heating by Law', *Solar & Wind Energy*, volume 3, 2008.

3. 'A generational challenge to repower America', speech by Al Gore, http://www.wecansolveit.org.

4. Ralf Gellings, Ines Rutschmann, 'And Don Quixote Wins in the End', *Photon Magazine*, January 2008.

5. Defined as on-site or decentralised energy production, such as waste-heat or gas-fired cogeneration, wind and solar power, geothermal, small hydro, and waste or biomass fuelled plants.

6. 'Using Energy More Efficiently: an Interview with Rocky Mountain Institute's Amory Lovins', The McKinsey Quarterly, 2 July 2008.

7. Ken Lewis, 'Markets alone will not lead to a green future', *Financial Times*, 6 June 2008.

8. Juliette Jowitt, 'Supermarkets come in from cold as part of low carbon revolution', *Guardian*, 25 October 2008.

9. Michael Rogol, 'The Black Swan: the Impact of the Highly Improbable', *Photon Magazine*, November 2008.

10. For detailed argument see Amory Lovins, E. Kyle Datta, Odd-Even Bustnes, Jonathan G. Koomey and Nathan J. Glasgow, *Winning the Oil Endgame: Innovation for Profit, Jobs, and Security*, Earthscan, 2004.

11. Aida Edemariam, 'The true cost of war', *Guardian*, 18 February 2008.

12. SIPRI Yearbook 2007: Armaments, Disarmament and International Security, Stockholm International Peace Research Institute, 2007.

13. Fiona Harvey, 'Green moves fall short of radical change', *Financial Times*, 13 March 2008.

14. 'Hot Air', *The Economist*, 15 March 2008: figures for 2005.

15. John Reed and Fiona Harvey, 'An industry charged up: electric vehicles are poised to go mainstream', *Financial Times*, 27 May 2008.

16. John Gapper, 'The humbling of Davos Man', *Financial Times*, 29 January 2009.

17. John Thornhill, 'A measure remodelled', *Financial Times*, 28 January 2009.

18. Paul Eccleston, 'Global warming "changing world economy"', *Daily Telegraph*, 8 January 2008.

19. Juliette Jowit, 'Green business boom is set to face trial by economic downturn', *Observer*, 25 May 2008.

Index

Numbers in italics indicate captions; those in bold indicate Figures

Abengoa 111, *112*
absorption cooling 89, 90
Abu Dhabi 124, 126
acid rain 136
Adams, William Grylls 57
adobe houses 9
Afghanistan, war in 129, 148
Agassi, Shai 122
agricultural sector, water pumps 136
agriculture
    dependence of agricultural sector on oil 23
    reduction in agricultural capacity 23
    residues 92
air conditioning xiii, 47, 49, 89, 91
air pollution 150
airflow 78
Akhenaten, pharaoh 6
Alberta, Canada: tar sands *29*
algae 13
Algeria: combined solar and gas (CSP) plant 111
Almaraz (Cáceres), Spain: solar farm *74*
alternating current (AC) electricity 9, 39, 40, 73
Amaterasu (Japanese sun goddess) 6–7
American Lung Association 150
amino acids, and 'primordial soup' 3
ammonia, and 'primordial soup' 3
Andasol plants, Spain 86, 112–23
Apollo (Greek god) 6
Apollo mission 143
Applied Materials 71
Arabian Gulf 123
Areva 32
arms-control agreements 149
Arup/Foster and Partners 24
Asda 146
Association for the Study of Peak Oil (ASPO) 24
asteroids 1
asthma 150
Aten (visible disk of the sun) 6
Athabasca Tar Sands, Canada 24
atmospheric pressure, and wind turbines 12
ATP molecule 3
Australia

coal reserves 19
coal-to-liquids plants 30
Austria: Kroiss 'plus-energy' house 105
Aztecs 6

Baden Würtemberg, Germany: solar PV *51*
Ban Kuai village, Laos 137
Bank of America 145
Barclays 129
baseload 50, 51
batteries 37, 39, 41–2, 45
    advanced 41, 44
    charging 41
    defined 41
    electrolyte 41–2
    lead-acid 41, 42
    lithium-ion 119, 120, 121
    recharge and battery-swapping stations 119, 122
    and stand-alone systems 9
    vanadium redox-flow 41–2
Bell Laboratories 57, 59
Bequerel, Edmond 57
Berlin climate summit (1995) 147
Berman, Elliot 58
Big Bang 1
biogas
    micro-combined-heat-and-power units 91
    powering combined-heat-and-power plants 37
biomass
    biomass-fuelled plants 39
    energy sources 12, 92
    as a renewable technology 36
birth rate 136
blackouts 41
Blood, David 132
boilers
    gas 10
    oil-fired 81
    pellet 81
boron atoms 59
Botswana: coal-to-liquids plants 30
BP 24
Brenninkmeijer, Marcel 132, *132*
Britain *see* United Kingdom
British Gas 100

bronchitis, chronic 150
buildings: transformation into solar power
    plants 103–8
Bureau of Land Management (US) 113
Bush, George W. 21, 148, 149
Bwelero village, Malawi *128*

C&A retail chain 132
cadmium telluride 65–6, 67, 71, 72
calendar stone, Aztec 6, *6*
California
    existing electricity capacity 113
    grid parity 117, **117**
    setting targets 144
    Social Initiative 143
    solar farms *120*
    candles 39
Cape Cod, Massachusetts: solar-powered
    home x-xiii, *xiii*
capitalism, redesigning 150, 153–4, **155**
carbohydrates, and primitive marine
    photosynthesisers 12
carbon
    emissions 18, 28, 35, 147, 154
    taxation 52
    zero-carbon objective 92
carbon capture 115, 131
    hydrogen generation 43
    and storage 30–31
carbon dioxide
    absorption of radiation 5
    biomass fuels 92
    carbon capture and storage 30
    cutting emissions: Woking's example xiii-xiv
    Earth's early atmosphere rich in 12
    electric car emissions 121
    energy use in buildings 103
    and fossil-fuel burning 20
    reduction in emissions 92, 133, 143
    UK retail chains' targets 146
carbon footprint of air conditioning 89
carbon monoxide production 43
Carboniferous period 13
cars *see* vehicles
Castile-La Mancha, Spain: energy mix xiv,
    144
Centrotherm 71
CFS (Co-operative Financial Services) 108
    tower, Manchester *107*, 108
Chernobyl nuclear disaster (1986) 32
childbirth, death during 136
China
    coal reserves 19
    coal-fired power plants xv
    coal-to-liquids plants 30
    and global warming assessment 22

lends to US 149
oil demand growth 26
public opinion on reliability of future oil
    supply 53
setting targets 144
solar corporations xvi
solar thermal collectors 107
solar water heaters xiv, 138
sun deities 6
Three Gorges Dam 126
chlorophyll 12
CIGS *see* copper indium gallium selenide
CIS (copper, indium, selenium) cells 66
Citibank 129
citizen development-financing 130
Clean Air Act (US) 150
Clean Development Mechanism (CDM)
    131
climate: historic stability 3
    climate change
    collective response to ix
    and countries' loss of wealth 154
    energy reform xii
Climate Change Committee (UK
    government) 114–15
climate crunch 20–23
Clinton, Bill 129
Co-operative group 145, *145*
coal
    beginning of large-scale coal
        burning 17
    coal-fired power plants 131
    costs 115
    dominates the resource tally 19
    electricity from 30
    extraction by unconventional ways 28
    far from capture and storage 30–31
    formation of coal seams 13
    meets over a quarter of global primary
        energy demand 17
    volume of remaining coal 14, *14*, 19
coal gas 79
coal reserves
    concentrated in the hands of very few
        countries 19
    size and national ownership 18, **18**
coal-fired power plants 44, 48
coal-to-liquids
    China halts coal-to-liquids plants 30
    an energy-intensive process 30
    and greenhouse gas 30
Code for Sustainable Homes 92, 104
Cold War 22
Colorado
    cliff-dwellers 9
    oil shales 29

combined-heat-and-power (CHP) plants
xiii, 37
combined-solar-and-gas (CSP) plants
111–12, *112*
comets 1
Concentrix *77*
copper 66
copper indium gallium selenide (CIGS) 65,
66, 67, 72
Cowan vs. Scargill ruling (1985) 133
crop growth 3
crystalline silicon 60–64, 74, 109, 115
cells to modules 63–4
grid parity 72
solar feedstock 60–61
wafers 61–2, *61*, *62*
wafers to cells 63, **63**
CSP plants *see* combined-solar-and-gas
plants
Cyprus: solar thermal per capita 107
Czochralski process 61

Day, Richard Evans 57
Denmark 149
Department of Business (UK) 31
depression (1930s) 98
Derby: solar facade system *153*
desalination 93–4, 123, 124, 135
Deutsche Bank 102, 129
Asset Management 132
Development Fund for Iraq 148
diesel, and greenhouse gas 30
diffuse radiation 5
dinosaurs 13
direct capital investment subsidies/rebates
142
'direct gain' system 88
direct-current (DC) electricity 9, 39–40, 73
doping process 59, 61, 63
'drainback' system 80
droughts 23, 123
DuPont 133
dust 5, 150
dye-sensitised solar cell 67, 72

Earth *1*
early atmosphere rich in carbon dioxide 12
life emerges on land 12–23
orbit around the sun 1, 4
primitive marine photosynthetisers 12
tectonic plates 13
tilt of its axis 4, **5**
economics ministries 50
Economist, The 52, 147–8
Eden Project, Cornwall: solar
architecture *141*

Edison, Thomas 39, 40, *40*
Edison Electric 40
education 135–8, *139*
Egyptians, ancient 6
Einstein, Albert 57
Eldorado, Nevada: solar farm *73*
electric grid systems 102
defined 37
history of electric grid systems 39–40
house 9
local xiv, **38**, 39
national xiv, 103
off-grid systems 37, **38**, 39
on-grid systems 37, **38**
'smart grids' 40, 46–8, **46**, 52, 103, 119–22,
144, 145
traditional 46, **46**, *48*
electric lighting 98
electric vehicles *see under* vehicles
electricity
alternating 9, 39, 40, 73
baseload 50, 51
central generation 37, **38**, 39, 40, 41, 46
direct-current 9, 39–40, 73
distributed generation 37, **38**, 39, 40, 41
from coal 30
generated in central power plants 37, 39
intermittency 41
meter reading 47
peaks of extreme demand 41, 45, 103
prices 79, 104, 116
solar-thermal power 10
stored energy *see* energy storage
uninterrupted power supplies 42
using off-peak electricity 45
utilities 145, 146–7
electrolysis: hydrogen generation 43
electrons 58–9
embedded generation 37
emphysema 150
encapsulation 64
'Energetikhaus 100', Germany 105
energy: annual global use *14*
energy conservation: replacing fossil-fuel
burning iv
energy crisis (1973) 82
energy efficiency
investing in 99–100
married with solar photovoltaics 92
maximisation in homes xii
replacing fossil-fuel burning ix
and solar thermal 92
energy payback 87
energy prices 100, 107
energy security x, 19, 131, 147
energy service companies (ESCOs) 98, 100

energy storage
advantages of 41
   batteries 41–2
   by 'smart buildings' 103
   hydrogen and fuel cells 43–4
   other forms of 44–5
enhanced oil recovery 25
entrepreneurs 132–3
Environment Agency HQ (UK): solar PV
   louvres *150*
environmental industries, global sales of 154
Environmental Industries Commission 154
environmental taxes 149
E.ON 31
EPIA *see* European Photovoltaic Industry
   Association
Equator Principles 133
Equity Bank, Kenya 133–4
ESCOs *see* energy service companies
ethical investment products 131–2
Ethiopia: first solar-powered village in 129
ethylene glycol 82
Europe
   GDP 153
   growth of solar thermal market xvi
   slow progress in reducing emissions/fossil
      fuel burning xiii
   smart-grid technology 48
European Commission 111
European Photovoltaic Industry Association
   (EPIA) 114
European Pressurised Water Reactor (EPR)
   32
European Social Investment Forum 132–3
European Union (EU)
   '20:20:20' climate deal (2008) 144
   carbon capture and storage 30
   Emissions Trading Scheme 144
   per capita energy consumption 35
Evergreen Solar 62
ExxonMobil 24, 29

feed-in tariffs 105, 107, 130, 141–2
financial crisis (2008) xii , 48, 85, 97, 118,
   131, 135, 150, 153, 154
firewood use, household 135, 136, 137
First Solar *64*, 66, 71, 115
Flammaville, France 32
flash distillation *see* vacuum distillation
floods 23
fluidised bed reactor 60, 87
flywheel 45
Fontana, California: solar installation *102*
forests
   first appearance of 13
   unsustainable forest harvesting 154

fossil fuels
   80% share of primary-energy supply 17–18
   building houses using no fossil fuels xii-xiii
   creation of greenhouse gases vi
   global reserves and resources 17–19
   replacing with clean energy 147
   retreat from dependence on xv
   strong case for switching away from 54
   subsidies 52, 128
   and the sun 12–24
fossil-fuel burning
   and carbon dioxide concentrations 20
   pollutants 149–50
   replacing with renewable-energy sources vii
fossil-fuel burning ix
fossil-fuel reserves, size and national
   ownership 18, *18*
Foster & Partners 124, *124*
France
   domestic feed-in tariff 107, 142
   nuclear power 32
Freshfield Bruckhaus Deringer 133
fridges
and built-in controllers 47
refrigerant cycle 89
solar 135
FT/Harris survey (2008) 53–4
fuel cells
   availability 43
   defined 43
   efficiency of 43
   and hydrogen 43–4
   'net energy' problem 44
   types of 43

G7 150
gallium 66, 76
gas
   beginning of widespread use 17
   costs 115
   meets over 20% of world primary energy
      demand 17
   natural *see* natural gas
   prices 75
   volume of remaining gas 14, *14*
gas boilers 10
gas fields, formation of 14
gas hydrates 18–19
gas industry, demand for finite products x
gas reserves
   concentrated in the hands of very few
      countries 19
   size and national ownership 18, **18**
gas-fired cogeneration 39
GDP *see* gross domestic product
gender imbalances 135, 136–7

General Electric 40
Generation III pressurized water reactors 32
Generation Investment 132
    Climate Solutions Fund 132
    Global Equity Strategy Fund 132
generators, diesel 42
geothermal heat pumps xiii, 36
geothermal power
    electricity provision 39
    use of natural heat from the ground 12, 35–6
germanium 76
Germany
    biggest solar-photovoltaic market xiii, 5
    'Combined Renewable Energy Power Plant'
        experiment 50–51
    domestic feed-in tariffs 105, 141
    efficient-energy technologies investment 101
    PV installation *101*
    solar homes *55*, 105
    solar industry xvi, 141
Ghana: microfinance initiative 129
Girl Children's Rights Association 138
Glasgow Science Centre: wind turbine *91*
global warming 147
    the maximum acceptable level 28
    projected level if emissions not cut deeply
        20, 21–2
GM EV1 battery car 121
Goldman Sachs 113, 132
Good Energies 132
Gore, Al 132, 143–4
Grameen Bank, Bangladesh 128, 133, *133*,
    134
Grand Inga Dam 126–7
Greeks, ancient 6, 9, 78
Green Energy for Schools scheme 145
'green new deal' 100, 102
greenhouse gases
    crop growth and water supplies in jeopardy 3
    emissions 21, 23, 123, 128, 133, 136, 149
    heat trapped in Earth's atmosphere ix, 20
    incoming radiation allowed through 20
    and oil depletion 28
    origin of ix
greenhouses 123
grid parity 68, 72, 73, 111, 114–18
grids *see* electric grid systems
gross domestic product (GDP) 153, **155**
    vs. population distribution **155**

health, healthcare 135, 136, 137, 149, 150
heart diseases 150
heat exchanger 88
heat pumps xiii 36, 89, 91–2
Helios (a Titan) 6
heliostats 49

helium
    and galaxy formation 1
    and stars 1
    and the sun 2
High Plains Ranch, USA 109
Hinduism 7
HIV infection 136
house-building
    mix-and-match approach xiii
    using no fossil fuels xii
hybrid systems 10, 131
hydrocarbons 150
hydroelectric power
    Congo River scheme 126
    and crystalline modules 87
    dependent on the sun 12
    using gravitational energy 44–5
hydrogen
    central creation of 43
    as a chemical energy-storage medium 43–4
    and fuel cells 43–4
    and galaxy formation 1
    the most abundant element in the universe
        43
    and 'primordial soup' 3
    and stars 1
    and the sun 2
    hydrological cycle 12
hydropower
    electricity generation 39
    as a renewable technology 36
    trebled by Portugal in under three years xiv
    unsustainable uses of 17

IGCC *see* Integrated Gas Combined Cycle
Imperial College of Science and Technology,
    London ix
India
    coal reserves 19
    coal-to-liquids plants 30
    oil demand growth 26
    public opinion on reliability of future oil
        supply 53
'indirect gain' system 88
indium 66, 76
Indonesia: coal-to-liquids plants 30
Industrial Revolution 17
Industry Taskforce on Peak Oil and Energy
    Security (ITPOES) 23, 25, 26
infrared radiation (R) 2
Institute for Energy (European
    Commission) 111
Institute for Public Policy Research 100
insulation 98–9, *99*
Integrated Gas Combined Cycle (IGCC)
    115

Intergovernmental Panel on Climate Change
(IPCC)
    four warnings by 20–23
    and grid parity 114
'intergrid' 48
intermittency 41
internal combustion engine 31
International Energy Agency 25, **25**, 26, 102,
128–9
    expected rate of oil demand growth 26
    'World Energy Outlook' (2008) 25, 29,
    117–18
inverter technology 9
investment 130–34, *132*, *133*
Iran: gas reserves 19
Iraq, war in 129, 148–9
ISAs 131
Islam 7
Israel: solar thermal in all new builds 143
Italy: feed-in tariffs 142

Japan
    coal-to-liquids plants 309
    flow-battery system used at a wind farm 42
    GDP 153
    plug-in hybrid vehicles 42
    setting targets 144
    solar industry xvi 141
    sun goddess 6–7
Jebel Ali desalination plant, United Arab
    Emirates 94
job creation 101, *101*, *102*
Johannesburg summit on development and
    environment (2002) 135

Kansas: wind power 36
Kashagan field, Caspian Sea 26–7
Kassel, Germany: solar building *35*
Kennedy, John F. 143
Kenya: kerosene lamp conversion 127
kerosene 39
    adaptation of lanterns to solar power xiii,
    126, 127–8, *128*, 129, 134, 136, 137
    dependency of developing world on kerosene
    for light at night xiii
kilowatt: defined 4
kilowatt-hours (kWh)
annual solar radiation 5, *5*
defined 4
kilowatt-hours per square metre 5, *5*
Kingsnorth, Kent 31
Kiva.org 134
Kohl's (retailer) 108
Konarka *67*, 68
KPMG 53, 149
Kroiss 'plus-energy' house, Austria 105

Kunming, China: solar water heaters 138
Kyoto Protocol 126, 131

La Paca (Alicante), Spain *111*
labour-intensive industries
    and the financial crash 85
    installation of solar photovoltaics and solar-
    thermal 85
lanterns *see under* kerosene
Lazard 115, **116**
lead-acid batteries 41, 42
LEDs *126*, 128, *128*
lender-philanthropists 134
Lewis, Kenneth D. 145–6
Lightning GT 121
literacy, female 136
lithium-ion batteries 42
local government 143
local grid systems xiv, **38**, 39
louvres, movable 9
Lovins, Amory 42, 148
low-temperature solar-thermal technologies
    78
lung diseases 150

McKinsey Global Institute (MGI) 99, 114
Malawi
    agricultural self-help 129
    kerosene lamp conversion 127
Malta 144
Manchester: solar PV clad CFS tower *107*,
    108
marine algae 13
Marstal, Denmark 81
Masdar Institute, Abu Dhabi 124, *124*
medical centres, solar-powered 136, 137
megawatt (MW): defined 4
Met Office: projected global warming if
    emissions not cut *20*
metallurgical-grade silicon 60
meteorite impacts 13
meter reading 47
methane
    emissions 20, 23
    and hydrogen 43
    and 'primordial soup' 3
'micro'-grid system **38**, 39
microcredit 128
microfinance movement 129
microloans 133–4
micropower technologies
combined-heat-and-power plants 37
    fast-growing use of 39
    in residential and commercial properties 103
    statistics of usage 145
microsolar entrepreneurs 129

Middle East
    desalination plants in 94
    proposed desert solar plants 111
military expenditure 148–9
military insurance 22
Milky Way 1
Mitsubishi Motors 121
mix-and-match approach xiii, 90–93
    air-source and ground-source heat pumps
        91–2
    biomass 92
    cost comparisons 92–3
    small- and medium-scale wind 91, 91
mobile-phone networks 137
modules
    cleaning 85
    crystalline 71, 87
    prices 69–70, 71
    production 64, 64, 69
    solar photovoltaic concentrator 77, 77
    solar-photovoltaic modules 103
Mojave Desert 113
moon, gravitational pull of 12
Morgan, JP 40, 129
Moura solar PV farm, Portugal 51–2
multicellular organisms 13
mutual funds 132

Nagpur, India 143
Nairobi Stock Exchange 133
nanocrystalline structures 65
Nanosolar 66
National Development and Reform
    Commission (China) 30
national grid xiv, 103
National Infrastructure Banks (proposed)
    102
natural gas
    discovery of resources 79
    formation of 14
    powering combined-heat-and-power plants
        37
NEC 120
Nellis Air Force base, Nevada 109
net metering 142
Netherlands 149
Nevada desert: solar thermal concentrators
    11, 82
new constituency, emergence and growth of
    145–8, 145, 146
new deal 98
'new economics' commission 153–4
New Zealand: coal-to-liquids plants 30
nickel metal hydride batteries 42
Nigeria: public opinion on reliability of
    future oil supply 53

nitrogen oxides 150
North Dakota: wind power 36
north pole: summer solstice 4
    npower 147
nuclear industry
    next generation of power plants 31
    Sizewell nuclear power station, Suffolk 33
    workforce 98
nuclear power 12, 31–3
    contributes 6% global energy demand 17
    Pinho's comment 52
    the sun 2

Obama, Barack 48, 143, 144
Oerlikon 71
off-grid systems 37, 38
oil
    annual consumption 17
    demand growth 26
    energy density 23
    era of 'easy oil' 24
    expansion of unconventional production 26
    first discovered 17
    formation of 13–14
    global supply vs. projected demand 24, 27, 27
    imports 49, 50, 143–4
    meets over 30of world primary energy
        demand 17
    pivotal to most processes in modern societies
        23
    public opinion on reliability of future oil
        supply 53
    reduction in discoveries 26–7
    unconventional oil resources 24
    volume of remaining oil 14, 14
oil crunch 23–8
    concerns about conventional oil 25–8, 25, 27
    peak-oil problem xii, 22, 23–5
oil industry
    demand for finite products xii
    exploration budget 27
    investment 26, 27, 29
    oil executives' opinions on renewable energy
        53
    profound infrastructure problems 27
    workforce 27, 98
oil prices 24, 26, 49, 53, 148
oil reserves
    concentrated in the hands of very few
        countries 19
    numbers possibly overinflated 19
    size and national ownership 18, 18
oil shales 29
oil-from-gas 26
oilfields
    depletion of production 25–6

enhanced oil recovery 25
  formation of 13–14
  old 24
  running out alarmingly fast 25, **25**
Olkiluoto 3 plant 32
Oman 123
on-grid systems 37, **38**
OPEC (Organisation of Petroleum
  Exporting Countries) 19, 24, 26
organic photovoltaics ('plastic PV') 68, *68*
oxygen: and primitive marine
  photosynthesisers 12
ozone: absorption of radiation 4–5

P & G 146
Pacific Gas and Electric 49, 74–5, 109, 111,
  113
parabolic trough solar concentrators 82, **83**
Pasadena, California 79
passive solar power **8**, 9, *9*, 78
pasteurisation 135, 137
peak oil xii, *22*, 23–5, 28
peer-to-peer lending 130
Pelamis 'sea-snake' wave generators, near
  Porto, Portugal 52
pension funds 130, 133
permafrost 23
Persian Gulf 143
Peru: solar cookers *136*
Philippines, coal-to-liquids plants 30
Phonexaysack, Dr Sonexay 137
phosphorus atoms 59
photosynthesis 2, 12
photovoltaic cells 57–60, **59**
photovoltaic effect 57–8, 64
photovoltaic vs. solar-thermal systems 84–7
  area required 85
  the condition of light required 84–5
  costs and prices 86
  energy payback 87
  water supply 86
  the workforce 85
photovoltaics value chain
  capturing the whole chain 70–72, *72*
  squeezing cost from the chain 68–70
Pinho, Manuel 52
pollutants, and burning fossil fuels 149–50
pollution: scattering radiation 5
polysilicon production 71
population distribution vs. GDP **155**
Portugal: clean-tech plan xii–xiv, 51–2
positive feedback effects 22
poverty alleviation 153
'power tower' solar concentrator 83, **83**
pregnancy, death during 136
'primordial soup' 3

Project Better Place 119–20, 122, *122*
proteins, as building blocks of life 3
protocrystalline structures 65
PS10 and PS20 plants, near Seville, Spain
  111
public transport 101, 102
Putin, Vladimir 150

Qatar: gas reserves 19

Ra (Egyptian sun god) 6
radiation
  annual solar *5*
  diffuse 5
  infrared (R) 2
  reduced by reflection, absorption and
    scattering 4
  reflected 5
  ultraviolet (UV) 2, 3
  visible 2
radioactive waste 31–3
Raishahi, Bangladesh: Shidhulai boat library
  139
REC 87
Red Army 22
reel-to-reel processing 67, *67*
reflected radiation 5
refrigerant cycle 89
Rema, Ethiopia 129
Renault-Nissan 119–20, *122*, 147
renewable energy
  connected to the grid 37
  contribution to global energy demand 17
  job creation 101
  in an off-grid system 37
  powering entire nations 49–52, *51*, **52**
  proportion of 'green new renewables' 17
  replacing fossil-fuel burning ix
  resource potential 36, **36**
  thirst for change 53–4
Renewable Energy Portfolio (Californian
  state government) 75
renewable portfolio standards 142
reserves
  defined 18
  and resources 18–19
resources
  defined 18
  depletion of non-renewable resources 154
  and reserves 18–19
respiratory allergies 150
retail chains 146
reverse osmosis 93, 94
Richmond, Surrey: UK's first solar roof 104,
  105
Rizhao, China: solar water heaters xvi, 138

Rogers, Jeffrey *xiii*
Romans, ancient 6
Rome 143
Rotherham, Yorkshire: solar roof tiles 107, *107*
runaway effect 22
running economies: the new priorities 148–50
Russia
    domestic oil and gas wealth xiii
    gas reserves 19
    oil reserves 19
    public opinion on reliability of future oil supply 53
    seen by Western Europeans as an unreliable energy supplier 53–4
    turns off Europe's gas supply briefly xvii

Safeway 108
Sahara Desert: proposed solar installations 50, 111
Sahara Forest Project 123
Sainsbury's 146
salt systems 45, 111–12, 113
San Francisco Bay Area: electric cars 121–2
San Luis Obispo County, California: solar farm 74, 109
sanitation 135
Sarkozy, Nicolas 153
satellites 58
Saudi Arabia: oil production 26
Saussure, Horace de 78
Scheer, Hermann xiii
Schlesinger, James 30
Schutzeichel, Harold 129
Schwarzenegger, Governor Arnold 144
Scientific American 50
Scottish and Southern Energy 24
sea-levels, rising 23
seasonal heat storage 45
seasons 4, **5**
seawater
    desalination 93–4
    evaporators 123
    nutrients from 124
seaweed 124
'seeing is believing' effect xvii
SEGS *see* Solar Energy Generating Systems
selenide 65
selenium 57, 66
semiconductors 9, 57, 58, 65, 84
Sempra Energy *73*
Sen, Amartya 153
Seville, Spain: solar PV *142*
Sharp 141
Shell 24, 35–6

Shidhulai Swanirvar Sangstha organisation, Bangladesh 137
    boat library, Raishahi *139*
Siemens process 60, 87
silane 60, 65
silicon
    amorphous 65, 74, 84, 109
    crystalline *see* crystalline silicon
    metallurgical-grade 60
    monocrystalline 61, *61*, 62
    polycrystalline 61, 62
    seeds 60
    solar-grade 60, 65
    thin-film 65
    upgraded metallurgical-grade (UMG) 68
silicon tetrachloride 60
Silicon Valley 52, 66, 134, 145
Sizewell nuclear power station, Suffolk *33*
smart grids *see under* electric grid systems
Socrates 78
sodium sulphur batteries 42
soil erosion 138
Sol Invictus 6
solar cells 9, *9*
    connected 64
    dye-sensitised 67, 72
    efficiency 63
    encapsulation 64
    expenditure on production 63, 66
    next-generation solar-cell technologies 67–8, *67*
    thin-film 64–6, **65**
    voltage 63–4
    wafers converted to 63
Solar Century 24
solar cookers *136*, 137
solar cooling 89–90, 124
solar economics
    costs of building a solar-powered house xiii
    recouping extra build costs xiii
    saving increasing amounts of money yearly xiii
Solar Energy Foundation 129
Solar Energy Generating Systems (SEGS) 82
solar generators, small 39
Solar Initiative (California) 143
Solar Investments 113
solar irradiance
    amount available for capture 8, **14**
    a measure of energy-flow per unit of area 2
    reduction of 4–5
    worldwide distribution **5**
solar markets, stimulating 141–3, *142*
solar observatory 7
solar panels
    achieving grid parity 73

concentrated PV panels 76
distributed generators 37
energy payback 87
fabs (factories for panel fabrication) 71
leasing of 146
Rizhao 138
solar photovoltaics (PV) **8**, 9, *9*
   in American south-west 50
   Cape Cod house xiii
   concentrators 10, *10*, 76–7, *76*, *77*, **77**
   electricity generation 37
   farms/plants xiii, 51–2, 71, 73–5, *73*, *74*, **75**,
     85, 86
   grid parity 68, 72, 73, 111, 114–18
   grid-connected systems 9
   growth of global market xvi
   organic photovoltaics ('plastic PV') 68, *68*
   peak power of panels 4
   photovoltaic cells 57–60, **59**
   photovoltaic effect 57–8
   and photovoltaic systems *see* photovoltaic vs.
     solar-thermal systems
   stand-alone systems 9
   thin-film photovoltaics 64–5, 71, 103, 115
solar photovoltaics industry
   investment 60
   rapid growth of 60, 69, 70
   reduced cost of solar energy production
     70–71
solar power
   electricity provision 39
   as a renewable technology 36
   and a renewables-powered future 35
Solar Power Corporation 58
solar prices 75
solar PV tiles *100*, 104, *105*, 107, *107*
solar radiation, annual *5*, *14*
solar space heating of buildings 88–9
solar stills 93
solar-advocacy business 50
Solar-Fabrik AG *72*
solar-grade silicon 60
solar-thermal power **8**, 10
   in American south-west 50, **52**
   collectors xi, 10, *10*, 78–83, **79**, *80*, *81*, *82*,
     83, 90, 103
   concentrated solar power 10, *11*, 82–3, *82*, **83**
   electricity generation 37
   farms/plants 45, 49–50, 82–5
   growth of solar-thermal market xvi
   low temperatures 78
   passive solar 78
   and photovoltaic systems *see* photovoltaic vs.
     solar-thermal systems
   solar space heating and cooling in buildings
     88–90

solar-thermal collectors 78–81, **79**, *80*, *81*,
   107
   water-heating 78–81, 105
   working with solar photovoltaics 90
SolarAid 127, *128*, *129*
   website (www.solar-aid.org) vi
Solarcentury x
solarisation 12, 35–7, 135, 138, 148
Solarworld 70
solstices 4, **5**
soot 150
South Africa 127
South Korea
   GDP 153
   tracker technology 75
south pole: summer solstice 4
South Yorkshire Housing Association
   107
Southern Africa 136
spacecraft 58
Spain
   Andasol plants 86, 112–23
   Castile-La Mancha energy mix xiv, 144
   CSP projects 111–12, *112*
   investment costs 86
   solar farms *xv*, 74, *74*, *111*
   solar PV in Seville *142*
   thermal and PV in some new construction
     and renovation 143
sports cars 120–21
Stagecoach 24
Standard Assessment Procedure (UK) 104
stars, transmission of energy 1
Stern Review (2006) 117
Stiglitz, Joseph 148, 153
   The Three Trillion Dollar War 148
Stonehenge 7, *7*
Subaru 121
subsistence farmers 128
sulphur oxides 150
summer solstice 4
sun 1–3, *3*
   creation of 1
   and fossil fuels 12–24
   gravitational pull 12
   high in summer/low in winter 9
   and hydroelectric power 12
   nuclear power 2
   size of 1–2
   temperatures 2
   and wave-power 12
   and wind turbines 12
   sun gods/goddesses 6–7, *6*
Sunderbans, Bengal 138
sundial 7
Sunpower 63

Suntech 70, 71
supercapacitor 45
Surya (Hindu solar deity) 7
Sweden 144

Talsiman 29
Tanzania: education 136
tar sands 18, 19, 24, 26, 28–9, *29*
target-setting 143–4
taxation
    exemptions 142
    incentives/credits 142
tectonic plates 13
Tenerife 123
Tenth Forum on Sustainable Energy
    (Barcelona, 2008) xv
terawatt-hours per year 4
terawatts
    defined 4
    global power requirement 8
Tesco 146, *146*
Tesla, Nikola 40, *40*
Tesla Roadster 120–21, *121*
Texas: wind power 36
Thatcher, Margaret (later Baroness) 20–21
thin-film photovoltaics 64–5, 71, 103, 115
    CIGS cells 66
    expected capacity 72
    factories 71
    grid parity 72
Three Mile Island disaster (1979) 32
tidal energy, solar component of 12
Tokyo Electric Power (TEPCO) 121
Tonatiuh (Aztec sun god) 6, *6*
Toyota 121
tracker companies 76–7
tracker technology 75
tracking dish solar concentrator 83, **83**
trains, electric 122
transport electrification 122
trichlorosilane 60
'triple junction' technology 65
Tropic of Cancer 4
Tropic of Capricorn 4
TVO 32
'20:20:20' climate deal (2008) 144

ultraviolet radiation (UV) 2
and ATP formation 3
underfloor cooling 90
Unisolar 65, 84
United Arab Emirates 124
United Kingdom
    ESCOs 100
    grid parity 117
    lack of a domestic feed-in tariff 105, 107

public opinion on reliability of future oil
    supply 53
setting targets 144
solar houses 104, *104*, 107, *107*
solar power for schools 145, *145*
United Nations (UN) 133, 148
    and greenhouse gas emissions 21
    and the IPCC 20
    World Climate Conference 21, 22
    United Nations Intergovernmental Panel on
        Climate Change 19
United States
    the biggest economic stimulus programme in
        US history 144
    borrowing from China 149
    coal reserves 19
    coal-to-liquids plants 30
    cost of Iraq war 148–9
    ESCO income 100
    GDP 153
    land grab for solar sites in south-west 113
    oil imports 49, 50
    refusal to ratify Kyoto Protocol 21
    renewable portfolio standards 142
    Republicans/Democrats attitudes to
        renewable energy 53
    slow progress in reducing emissions/fossil
        fuel burning xiii
    small-grid technology 48
    solar-photovoltaic farms 50
    solar-thermal energy 50, **52**
    unemployment 144
    wind power 36
University of New South Wales 63
upgraded metallurgical-grade silicon
    (UMG) 68
US Department of Energy 87
US National Renewable Energy Laboratory
    66

vaccinations 135, 137
vacuum distillation (flash distillation) 93–4
vanadium redox-flow batteries 41–2
vehicles
    electric (EVs) 30, 32, 42, 47, 48, 119–22, *120*,
        121, *122*, 145, 147, 150
    hybrid 42, 119, 121, 122, 145
    and job creation 101
    petrol 119, *120*
    recharge and battery-swapping stations 119,
        122
    zero-emission 147
ventilation 98
vertical integration 70
Virgin 24
visible radiation, allows photosynthesis 2

visions of solar in action
    carbon armies rebuilding economies after the
        great financial crisis 97–102
    energy storage plants throughout smart grids
        119–22
    energy utilities' power generation 109–113
    more ambitious UN development goals
        135–8
    most buildings transforming into solar power
        plants 103–8
    the rural poor reading at night 126–9
    socially responsible investment 130–34
    solar farms and agricultural renaissance
        123–4, 126
    solar photovoltaics at grid parity 114–18

wafers 61–2
    conversion to solar cells 63
    manufacture 69
    mono-crystalline 61
    polycrystalline (multicrystalline) 61
Wal-Mart 108
waste-fuelled plants 39
waste-heat cogeneration 39
water
    groundwater 136
    and hydrogen 43
    and photovoltaics 86
    and 'primordial soup' 3
    quality of 135
    reliance on water supplies 3
    shrinking supply of 123
    solar pumps 135–6
    and solar-thermal power plants 86
    sterilisation 136
water vapour: absorption of radiation 4–5
water-heating devices 78–9

watt: defined 4
wave-power
    Portuguese wave-power station xv
    as a renewable technology 36
    and the sun 12
Wen Jiabao 150
Westinghouse, George 40
wind power
    electricity provision 39
    quadrupled by Portugal in under three years
        xv
    as a renewable technology 36
wind turbines 91, *91*
and atmospheric pressure 12
and the sun 12
wind 'wake' 91
wind-power stations (wind farms) 36, 42, 51
windows
    adjustable smart 9
    generation of solar electricity 103
winter solstice 4
Woking, Surrey xiii
wood, unsustainable uses of 17
Woolsey, Jim 147
World Bank 154
World Climate Conference 21, 22
World Economic Forum (2009) 150
World Future Energy Summit (2009) 126
World Health Organisation 150
WorldPublicOpinion.org 53
Worldwatch 'State of the World' report 133
Wyoming oil shales 29

Xinhua News Agency 30

Yahoo 24
Yunus, Muhammad 133, *133*